THE HISTORY AND USE
OF OUR EARTH'S
CHEMICAL ELEMENTS

THE HISTORY
AND USE OF
OUR EARTH'S
CHEMICAL ELEMENTS

A Reference Guide

ROBERT E. KREBS

ILLUSTRATIONS BY RAE DÉJUR

GREENWOOD PRESS
Westport, Connecticut • London

Library of Congress Cataloging-in-Publication Data

Krebs, Robert E., 1922–
    The history and use of our earth's chemical elements : a reference
guide / Robert E. Krebs.
        p.   cm.
    Includes bibliographical references (p.   ) and index.
    ISBN 0–313–30123–9 (alk. paper)
    1. Chemical elements.   I. Title.
QD466.K69   1998
546—dc21   96–49735

British Library Cataloguing in Publication Data is available.

Library of Congress Catalog Card Number: 96–49735
ISBN: 0–313–30123–9

First published in 1998

Greenwood Press, 88 Post Road West, Westport, CT 06881
An imprint of Greenwood Publishing Group, Inc.

Printed in the United States of America

The paper used in this book complies with the
Permanent Paper Standard issued by the National
Information Standards Organization (Z39.48–1984).

10 9 8 7 6 5 4 3 2 1

CONTENTS

INTRODUCTION

This volume is a reference for students and other readers, as well as for school and public library use.

It introduces the background of how we came to know and understand the **chemical*** nature of our Earth and everything in the universe, including ourselves. Early people had limited knowledge and use of their chemical environment. Through the ages they progressed from a practical to a rational approach to the nature of the chemical **elements** found on Earth.

As science developed, our accumulation of knowledge of the structure of **atoms** and **molecules** was an achievement of early philosophers, thinkers, and scientists. These men and women built the foundation of our current understanding of the structure of **matter**. This history has led to our current understanding of the theoretical and practical nature of the chemical elements.

This work describes the chemical elements according to characteristics such as structure, size, weight, activity, abundance, usefulness, and hazards. Each element's structure relates to its "fit" within the Periodic Table of the Chemical Elements. The book is about the chemical elements found on Earth as well as in the entire universe. Interestingly, the proportions (by weight or numbers) of the elements to each other found on Earth are not the same ratios as found in the universe. While hydrogen is the most common

*Terms in bold type can be found in the Glossary of Technical Terms.

element in the universe (90%), the most common element on Earth is oxygen (49%).

Chemistry is a physical science that studies the structure and properties of elementary matter and how that matter interacts with other substances. Matter can be defined as something that occupies space, has mass, and can be perceived by one or more of our senses, as opposed to something like the mind or spirit. We usually think of matter as the chemical elements composed of atoms, and compounds composed of molecules, which are combinations of atoms. Chemistry is the science of how and what happens when two or more atoms combine to form myriad **compounds** that make up all physical things we can see, touch, and taste. Chemistry is universal. We study the chemical interactions of atoms and molecules in the stars, the Earth's matter, living **cells**, and everything that exists. Chemistry is also a science of energy. The formation of molecules by the combination of atoms involves energy. Some **chemical reactions** (the combining of atoms) release energy; others require an input of energy to complete the reaction. The electrons in orbits surrounding the nuclei of atoms exist in a specific state or level of energy. Atomic structure is discussed in Chapter 2.

Almost everything we live with is made up of about one hundred different chemical elements that, in various combinations, form compounds that make up the Earth. The air we breathe, the food we eat, the clothing we wear, our bodies, and just about everything else we can think of consists of chemicals. All the molecules in our bodies are made up of six major elements: carbon, oxygen, hydrogen, nitrogen, phosphorus, and sulfur. There are also "trace" elements important to our well-being. Some of these are copper, iron, chromium, magnesium, calcium, and zinc. The current market price for all the elements in your body is about $9.95.

What makes chemistry so interesting is that each specific chemical element is related to its own kind of atom. Elements with specific characteristics have unique atoms. Each type of atom is unique to that element. If you change the basic structure of an atom, you change the structure and properties of the element related to that atom. Also of interest is what happens when two or more different atoms combine to form a molecule of a new substance. The original atoms no longer exhibit any of their original properties once they form a molecule of a new compound.

All this is dependent on the electron arrangement in the **shells*** around the nucleus of the atom. How atoms interact, that is, combine to form a

---

*The illustrations that depict the electron configurations for the atoms of each element are based on the Bohr model of energy shells.

molecule, is dependent on the arrangement of their electrons. The electron is one of the three major **subatomic particles**. It has a negative charge and a negligible **mass** (only 1/1837 the weight of a proton). Although electrons are continually circling the central **nucleus** of the atom, it is not possible to determine exactly where they are within their orbit (shell) at any point in time. The outer ring (shell) of electrons consists of a specific number of electrons for each element. For most elements these outer electrons may be thought of as **valence** electrons because they determine the chemical properties of atoms and how they combine with each other. Valence is "a whole number that represents the combining power of one atom to another, and provides the relative amounts of each of the interacting elements in the new molecule, i.e., compound" (Lewis, 1993). We will be referring to valence later.

This book uses the Periodic Table as the basis for organizing the presentations of the elements. Once you learn how this remarkable chart is organized, you will be able to relate the characteristics of many elements to each other based on their structure, which can be determined by their placement in the Periodic Table.

The first three chapters set the stage for use of the reference sections, which include the descriptions of the elements. Chapter 1 starts with a short history of science and chemistry, progressing from prehistoric times to the Age of Alchemy, and then to the Age of Modern Chemistry. Chapter 2 traces the history of our knowledge of the structure of the atom, some theoretical models, nuclear radiation, and chemical bonding. Chapter 3 introduces the Periodic Table and describes how to use it.

Chapters 4 through 10 discuss each element in detail. Of particular importance is that the presentation for each element is based on that element's position within a **Group** in the Periodic Table. Each element is presented as a separate entry which describes: 1) the element's atomic weight, atomic number, the period where it is found in the Periodic Table, its natural state, and its common isotopes (if any); 2) the important chemical and physical properties of each element; 3) the electron configuration for the atoms of each element; 4) each element's abundance and its source on the Earth; 5) the history of each element, which includes who discovered the element and how and when it was discovered; 6) examples of common uses for each element; 7) some examples of important molecules (compounds) for each element; and 8) each element's potential hazards to humans and the environment (if any).

At the end of the book there is a Glossary of Technical Terms, an Alphabetical Index of the Elements, an Electron Configuration of the Atoms, a Selected Bibliography, and an Index.

1

A SHORT HISTORY OF CHEMISTRY

THE BEGINNING OF SCIENCE

There are two major theories of how science developed. One states that early mankind, being curious and having some intelligence, began to explore nature by using trial and error. To exist, humans learned what to eat, how to protect themselves, and, when time permitted, how to cope with the environment to make life easier and more understandable. This is the "continuum" or "accumulative" approach to science and discovery—which is still going on.

The other theory, as presented by Alan Cromer in *Uncommon Sense—The Heretical Nature of Science* (Cromer, 1993), postulates that science was not a natural sequence of invention and discovery from ancient to modern times. He states that science, as we think of it, developed in early Greece, possibly because of the democratic nature of its culture, which included objective inquiry and debate. From Greece, science spread to China and India. Then the Arab world, as it conquered many countries, introduced science through southern Europe and as far west as Spain. Science was developed further in Greece and Egypt, and finally was brought to central and western Europe.

The libraries of Alexandria (North Africa) became depositories for people's knowledge and were major contributors to the advancement of science in many other countries.

Modern science is very different from the descriptions of early systems of thought. Early philosophers and thinkers lacked the objective methodologies and processes required for controlled experiments that led to modern science. They were more concerned with seeking universal cures for sickness, **transmutation** of **base metals** to gold, and mysticism in general. Most, but not all, depended more on the written words of "experts" than on their own observations and insights.

ORIGIN OF THE EARTH'S CHEMICAL ELEMENTS

We really do not know how the universe and our Earth and the elements came into being. It is believed to have all started over 20 billion years ago. Man has always espoused theories as to the origins of the universe—including philosophical, scientific, and theological theories, as well as just plain speculation.

One is the big bang theory, which states that it all started with the "explosion" of an incredibly small, dense, and compact ball of matter that rapidly expanded. Our observations indicate that the universe is still expanding, and this original unknown primordial mass is the source of all the chemical elements and energy existing in our universe. The formation of the chemical elements began in just seconds. Hydrogen, which has only one **proton** in its nucleus, was the first element. Hydrogen is still the most abundant element in the universe. In the heat of the explosion, hydrogen nuclei fused to form helium atoms, each of which has two protons in its nucleus. Together, hydrogen and helium make up 98 to 99% of all the atoms in the universe. As these gaseous elements condensed, they formed galaxies of stars with accompanying planets, comets, meteors, and all the chemical elements, including cosmic dust and junk.

Another theory speculates that there was no beginning and there will be no end to the universe—it is infinite. There is continuous death and rebirth of stars and matter in the universe. Others postulate that it all started by some spontaneous and unknown force.

Obviously, there are other theories that could be suggested, but they may require some unseen, unknown, all-powerful force or entity to get it all started. The origin of our universe is one of the big questions for which scientists, as well as philosophers and theologians, will continue to seek answers.

We do know a great deal about the nature of the universe. For instance, the element hydrogen makes up about 75% of all the **mass** in the universe. In terms of **number**, about 90% of all atoms in the universe are hydrogen atoms, while most of the rest of the atoms in the universe are helium. All the other elements in the universe make up just one or two percent of the total.

Interestingly, the most abundant element on Earth (in number of atoms) is oxygen. Oxygen accounts for about 50% of all the elements found on Earth. The second most abundant is silicone. Silicone dioxide (SiO_2) (sand and rock) forms about 87% of the Earth's mass.

EARLY USE OF CHEMISTRY

Early understanding and use of chemistry and the elements are not well recorded. Obviously, people did learn, one way or the other, how to use the Earth's chemical elements for survival. One of the earliest uses of chemistry was fire. People knew that it gave light and heat and, in time, learned how to use fire to their advantage, most probably for cooking. Sometime later, they discovered how to develop an oil lamp using liquid fat, and then solid fat to form candles for light (about 3000 B.C.). A great deal of chemistry is involved with a burning candle, but it took many centuries for mankind to understand it. An interesting application of early chemistry was the use of fire to make pottery from clay pots.

Early pots were formed from soft clay (circa 9000 B.C.) that could not hold much weight or water. Early people did have baskets and wood containers—yet there was a need for something in which to carry water and to cook food. About 7000 B.C. people learned, either by accident or by trial and error, to use fire to bake the soft clay, which then became hard. Once fired, the pot became a ceramic; thus a more useful container was developed. This may have been the first time mankind used fire for a purpose other than for cooking, heat and protection.

Another example of early use of chemistry is the discovery of **fermentation**. A very early discovery involved what occurred when fruit juices, left uncovered, turned into wine. During fermentation large glucose molecules (sugar) break down into the smaller molecules of carbon dioxide and alcohol. In June 1996, the Associated Press reported that wine-making residue was found at the bottom of a 7000-year-old clay pot. Chemical traces of wine were found in a jar that was located in the mountains of Iran. The previous recorded date was about 1800 B.C. Wine making is an old, well-known chemical process.

A related discovery, possibly an accident, happened when someone noticed that when dough made from flour that was pounded out of wild grains was exposed to the air, it "rose," forming leavened bread. Dough activated by yeasts, and then baked, is more edible because of the open texture produced by the gas carbon dioxide (CO_2), a by-product of fermentation. It became a practice to save a portion of one batch of dough as a "starter" for future baking. A starter batch of dough contains a small portion of the yeast used in the previous batch. It was kept separate and cool until mixed with the dough for the next batch of bread. Then a new starter was put aside for future baking. If you heat or bake a starter sample, you run the risk of killing the yeast, which is exactly what happens after you bake a loaf of bread that has been raised by action of the carbon dioxide.

There are numerous examples of the early uses of chemistry in metal-working. Copper and other metals normally exist in nugget form. About 4000 B.C. it was discovered, probably by accident, that when copper ore is heated it combines with air to form CO_2 and metallic copper (Cu). This may have been the first time that mankind did not need to rely on stone and wood for tools and weapons. A discovery was made (circa 3600 B.C.) that if you mix copper ore with tin ore, a tin-copper **alloy** called bronze results. This was the beginning of the *bronze age*. Bronze is stronger and holds a sharper edge on tools and weapons than does copper.

High-grade iron exists in meteorites and in some iron ores. Early mankind found these sources of iron, but could not do much with these chunks of more or less pure iron. Since most iron on Earth exists in ores where it is combined with other substances, it cannot be melted down by using wood fires, which were just not hot enough. In about 1500 B.C., humans learned how to convert wood to charcoal (another chemical reaction) and found that charcoal produces a higher temperature than wood when burned. This higher heat made it possible to **smelt** the iron from its ore. Iron made sharper, stronger, and more desirable tools and weapons than did bronze, thus the beginning of the *iron age*.

These examples, as well as other early uses of "chemistry," all involved chemical changes that are well known today. It was many years before humans began studying how chemical changes occur, and how to explain and control these reactions.

In 340 B.C. Aristotle (384–322 B.C.) published *Meteorologica*, in which he postulated that the Earth's matter is composed of four elements—earth, water, air, and fire. His speculations led to the idea that the Earth is composed of "shells." This was the extent of mankind's understanding of the composition of the Earth's chemical elements for several centuries—until the *age of alchemy*.

THE AGE OF ALCHEMY

The age of alchemy is said to span from about 500 B.C. into the seventeenth century. It was not until about 320 B.C., at the time of Alexander the

Great, that alchemists made serious studies of chemical changes. Alchemy encompassed Greek and Egyptian as well as Arabian concepts of matter and energy. The early alchemists were not very "scientific," as we think of the term today. They did try to figure out how different chemicals reacted, and they did make some discoveries that advanced knowledge. In the year 300, Zosimus (240–?) of Egypt made the first attempt to summarize the knowledge of the alchemists. Unfortunately, most early alchemists are unknown because they did not leave much written history.

Their goals were mystical and economic and were kept secret. Their practices also related to medicine, and during some periods of time and in some countries, alchemy was related to various religions. Alchemists were not scientists as we now think of the profession. Their main search was for the "philosopher's stone" which could unlock the secrets of how to transform base metals and chemicals into different, more useful products, such as gold and silver. This also led to the futile search over many centuries for the secret elixir that would be the universal "cure" for all illnesses.

For over 2000 years, alchemy was the only "chemistry" studied. As the predecessor of modern chemistry, alchemy was part of the slow growth of what we now know about the Earth's chemical elements. For example, the alchemists' interest in a common treatment for all diseases led to the scientific basis for the art of modern medicine. And their elementary understanding of how different substances react with each other led to the concepts of atoms and their interactions to form compounds.

THE AGE OF MODERN CHEMISTRY

It is difficult to determine an exact date for the beginning of modern chemistry. Some date the end of alchemy and the beginning of modern chemistry to the early seventeenth century. It is also impracticable to say just who is the "father" of modern chemistry. Many people, men and women of many races and countries, contributed to our current knowledge and understanding of chemistry over many years. A few examples:

In 1661 Robert Boyle (1627–1691) published a book titled *The Skeptical Chymist*, which was the beginning of the death of alchemy. His book ruled the thinking of early scientists for almost 100 years. Two of his contributions were the use of experimentation to determine properties of the chemical elements, and the concept that an element is a substance that cannot be changed into something more simple. Robert Boyle is best known for Boyle's law, which states that the volume of a gas varies inversely with the pressure applied to the gas. In other words, as you squeeze a container of gas, the volume of the gas decreases. In the reverse, if you increase the amount of a gas in a closed container, the pressure on the container becomes greater. In the early 1700s, Georg Ernst Stahl (1660–1734), a German chemist, developed a theory that when something burned, *phlogiston* (from the Greek "to set on fire") was involved. His idea was that burnable things had a limited amount of phlogiston, and that when burned, they lost their phlogiston, leaving residues that would not burn because they no longer had phlogiston. This theory did not hold up very long. Through "experimentation," it was shown that different products resulted from **combustion**, depending on what was burning.

Most historians consider that the French chemist Antoine-Laurent Lavoisier (1743–1794) completed the death of alchemy and gave birth to modern chemistry. He made many contributions to chemistry, including the important concept that in addition to just observing something, one also must make measurements of what one is observing and keep accurate written records. Lavoisier mixed substances, burned common materials, and weighed and measured the results. His work led to the discovery of over thirty elements. He described **acids**, **bases**, and **salts** as well as many organic compounds. By a unique experiment with water (H_2O), he determined that it is made up of the gases hydrogen and oxygen. From these experiments Lavoisier determined that oxygen has a weight eight times that of hydrogen. This led to a later theory of the Law of Definite Proportions, which states that a definite weight of one element always combines with a definite weight of the other in a compound. (NOTE: Lavoisier was not aware that 2 atoms of hydrogen combine with 1 atom of oxygen to form water. Therefore, the actual ratio of weight (atomic mass) is 1:16 instead of 1:8.) Until this time no standard nomenclature (names and symbols) was used for the elements, compounds, etc. In 1769, Lavoisier and others published a book titled *The Methods of Chemical Nomenclature*, which proposed a logical, systematic language of chemistry. With modifications by the Geneva System of 1892, and additional reforms by the International Union of Pure and Applied Chemistry (IUPAC) in 1930, Lavoisier's nomenclature of chemical names and symbols is still in use today.

Jöns Jakob Berzelius (1779–1848), a Swedish chemist, is also considered one of the founders of modern chemistry. He prepared, purified, and identified over 2,000 chemical elements and compounds. He determined the atomic weight (mass) of several elements and replaced pictures of the elements with symbols and numbers, which is the basis of our chemical notations today.

ELEMENTS VS. COMPOUNDS

We have used both the terms **weight** and **mass**. For most of us on Earth we can use weight and mass interchangeably, but there is an important scientific difference. The mass of an object is the amount of matter contained in a particular body, regardless of its location in the universe. The mass of an object is constant and always the same wherever that particular body is located, no matter what planet it is on.

The weight of an object relates to its size and distance from the gravitational pull of a larger body such as the Earth (or any other large body). In other words, weight is gravity's effect on an object. Thus object A's weight depends on two factors: first, the size (mass) of the two bodies (object A and the Earth); and second, the square of the distance separating the two bodies. Mass is constant in the universe, while weight may be thought of as the strength with which gravity "pulls" objects to the Earth.

The concept of an "atom" is very old. About 440 B.C., Democritus (ca. 460–370 B.C), a student of the Greek philosopher Leucippus, first proposed the concept of atoms. Democritus proclaimed that all matter is composed of tiny **particles**, and that no smaller particles can be separated from this indivisible "atom." Democritus had no experimental evidence, but speculations led to the atomistic view of matter that was not well accepted at that time, but prevailed for many centuries.

As mentioned previously, many people made contributions to modern chemistry. An important advancement in our understanding of Democritus' concept of the indivisible atom occurred in 1808 when the English chemist John Dalton (1766–1844), after conducting many experiments, published his book *New System of Chemical Philosophy*. In essence, he said that the atoms of each element are exactly alike and have the same weight; that the atoms of each element are different from the atoms of every other element, and that combinations of the elements is merely combination of the atoms in simple or multiple units. At this time Dalton did not know the exact number of atoms that could combine to form molecules. But, by experimentation he knew the relative weights of the elements forming compounds. From this information he developed the first table of **atomic weights**. His table was not accurate, but it was another step in the advancement of our knowledge of chemistry.

It became clear that atoms were the smallest, indivisible units of elementary substances, but it was important to find out the atomic weights and sizes of different atoms. Since atoms are too small to see with a microscope, counting and weighing them, as we do with larger objects, is impossible. Therefore, we need a different standard, and this is where the concept of "relative weights" becomes useful. It is possible to determine the mass of an individual atom relative to the mass of another atom. With this system to determine atomic weight (mass), chemistry advanced rapidly. In a sense, all standards used for measurements of weight (pound and kilogram), distance (mile and kilometer), volume (gallon and liter), and so forth are "related" to something.

Combinations of *similar* atoms form molecules, and combinations of atoms of *different* elements form compounds. Thus, there are at least two distinct types of molecule. One is a molecule of an element where one or more atoms of the same element combine to form a new molecule. For example, two oxygen atoms combine to form a molecule of oxygen gas (O_2). In the other type of molecule, atoms of different elements form molecules of new compounds; e.g., sodium and chlorine combine to form a molecule of salt (NaCl). In both types of molecules, the properties and characteristics of the molecules are different from those of the original elements.

Based on what we have learned, let's define elements and compounds.

ELEMENT: The smallest unit of each of the 112 different basic types of matter that either exist in nature or are artificially made. All atoms that compose a specific element are the same in nuclear charge and in number of electrons and protons. Atoms of some elements may differ in mass when the number of **neutrons** in that atom's nucleus differ (called **isotopes**). This will be discussed in the next chapter.

COMPOUND: A molecular substance composed of atoms formed by a chemical reaction of two or more elements. The molecules of the new compound have properties very different from the properties of the elements that formed the compound. An example is when hydrogen gas ($2H_2$) combines with oxygen gas (O_2) to form water molecules ($2H_2O$), which has none of the properties of the original two substances.

ALLOTROPE: An element or compound that exists in more than one form. Carbon is an example of an element found in different forms, e.g., carbon black, graphite, and diamonds. Oxygen has three allotropes: **monatomic** oxygen (O), **diatomic** oxygen (O_2), the gas we breathe, and **triatomic** oxygen (O_3), which is known as ozone.

2

ATOMIC STRUCTURE

EARLY IDEAS OF ATOMIC STRUCTURE

As we saw in Chapter 1, many scientists made important contributions to chemistry. They identified chemicals, determined their characteristics, similarities, and differences, and designed symbols for them. Using unique experiments, scientists devised ways to define the structure of atoms and determine atomic weights, sizes, and electrical charges, as well as energy levels for atoms.

Many of these men and women recognized the existence of some order in the way chemicals relate and react to each other, and that the structure of each element's atoms has something to do with these characteristics. There were several attempts to organize the elements into a chart that reflected the particular nature of the atoms for these elements. Before the Periodic Table of the Chemical Elements, as we know it today, was developed, several relationships had to be established.

The concept of "electrons" was known for many years, but how these negatively charged particles react required exploration. In about 1897, Joseph John Thomson (1856–1940) sent streams of electrons through **magnetic fields**. A dispersion of the electrons occurred. These experiments, and those of others, led Thomson to speculate that the atom was a positively charged "core" and that negatively charged particles surrounded and matched the positive charge of this core or nucleus. And further, when these electrons were excited or "stirred up" with a strong light, electricity, or magnetism, some of them were driven from the atom. This was one of the first experimental explanations for the structure of the atom. Many refinements of this concept were made by the work of several scientists, including the French chemist Marie Sklodowska Curie (1867–1934), the British physicist Ernest Rutherford (1871–1937), the American physicist Robert Andrews Millikan (1868–1953), and others of many nationalities.

There were still some questions that required answers before a logical organizational chart of the elements could be completed, e.g., how to determine the energy, position, and number of electrons in atoms of different elements. When electrons are "excited" by an input of energy, they jump to a higher energy level, orbit, or shell and then return to their original energy level, orbital position, or shell. When they return to their original position, they emit unique electromagnetic frequencies or light. This is true for the atoms of all the elements. This release of energy occurs in very uniform "packages" of energy. Each unit of energy is called a **quantum**, a small bit that cannot be divided into smaller bits.

Max Karl Ernst Ludwig Planck (1858–1947), following some theories of electromagnetic radiation, devised a mathematical equation to explain the relationship between energy and wavelengths of light. We now refer to this as Planck's constant, which is a very small measurement of a very small change in energy. The concept is so fundamental that it heralded the beginning of modern physics in the early 1900s.

Werner Karl Heisenberg (1901–1976), the founder of quantum mechanics, described the behavior of very small particles and their interactions based on some higher mathematical concepts. The theory is somewhat more easy to understand in a descriptive sense. His most unusual theory changed classical physics and is now referred to as the "uncertainty principle." In essence, it states that it is impossible to know both the exact position and mass of a subatomic particle at the same moment in time, but an accurate prediction (average) is possible. An interesting aspect of the Heisenberg uncertainty principle is that just viewing and measuring subatomic particles may alter their positions and other characteristics.

Wolfgang Pauli (1900–1958), an American physicist, was awarded a Nobel Prize in 1945 for developing the "exclusion principle." In essence, it states that a particular electron in an atom has only one of four energy states, and that all other electrons are excluded from this energy level. In other words, no two electrons may occupy the same state of energy. This led to the concept that only a certain number of electrons can occupy the same shell or orbit. In addition, the wave properties of electrons are measured in quantum amounts and are related to the physical and, thus, chemical properties of atoms. These concepts enabled scientists to precisely define important physical properties of the atoms of different elements and to more accurately place elements in the Periodic Table.

The terms **shell, orbit**, and **energy level** are sometimes used interchangeably. We will be using the term "shells" most often, because it is descriptive and conveys the image of three-dimensional layers or structures surrounding the nucleus. The term "orbit" is more closely related to the image of two-dimensional concentric rings, similar to an archery target. The term "energy level" is used to describe the energy interactions of atoms. The term "orbital," or rather the more descriptive term "suborbital," will be used later to describe the four distinct energy levels described in Pauli's exclusion principle. These four energy levels and the maximum number of electrons that can exist in each suborbital are:

1. The first suborbital energy level, with a maximum of 2 electrons, is called "s."
2. The second suborbital energy level, with a maximum of 6 electrons, is called "p."
3. The third suborbital energy level, with a maximum of 10 electrons, is called "d."
4. The fourth, and final, suborbital level, with a maximum of 14 electrons, is called "f." This concept will be discussed later.

SOME THEORETICAL ATOMIC MODELS

The earliest concept of atomic structure goes back to Greece in the fifth century B.C. when Leucippus and Democritus postulated that tiny particles of matter, which they called atoms, were indivisible.

Over the centuries many other concepts were proposed to explain the nature of matter. It was not until J. J. Thomson (1856–1940) proposed his model of the atom, which consisted of a sphere with a bunch of negative particles somehow positioned in a random fashion inside a very small ball of matter, that the modern structure of the atom began to take shape.

The Rutherford model of the atom is a significant improvement over the Thomson model. Sir Ernest Rutherford (1871–1937) incorporated the background and understanding of many scientists as he developed experiments designed to show that the atom has a central, small, but very heavy, nucleus. His experiments verified that this positively charged dense nucleus has negatively charged electrons surrounding it. This concept resembles the planets revolving around the sun, including the laws of motion and energy.

The Bohr model of the atom took shape in 1913. Niels Bohr (1885–1962), a Danish physicist, started with the classic Rutherford model and applied the new theory of **quantum mechanics** to develop a new model that is still in use. His assumptions are based on several aspects of the quantum theory. One assumption is that light is emitted in tiny bunches of energy called photons (quanta of light energy).

The Bohr model continues:

First, the orbital or quantum theory of matter assumes that the electron is not a particle, as we normally think of particles. **Orbital** theory considers the electron as a three-dimensional wave that can exist as several energy levels (orbitals).

Second, the electrons are in constant motion around the nucleus, even though it is not easy to determine the position of a particular electron in its shell at any particular moment.

Third, the electrons are revolving at different distances from the nucleus, called shells, orbitals, or energy levels.

Fourth, there are more than a few shells, orbitals, or energy levels.

Fifth, the electrons can continue to move in a specific shell without emitting or absorbing energy.

Sixth, the electron will remain in the shell closest to the nucleus or to the lowest energy level that it can maintain.

Seventh, if the electron is excited by external energy, it can "jump" to a different or higher energy level (shell).

Eighth, when the electron returns to its former shell or lower energy level, it will emit the energy that it acquired to raise it to the higher level or orbit.

Bohr's ideas led to the structure of the atom being compared to the structure of an onion. The outer layers of skin on an onion are the "shells" where the electrons exist. At the center of the onion is a dense, tiny "BB-like shot" for a nucleus. The area between the nucleus and the electrons' shells is just space. There is no distinct boundary for the shells. The electrons assigned to a particular shell are in motion all the time, so the shell does not seem to have a sharp definition.

The "Fuzzy Ball" Atom

It is something like a "fuzzy ball," with no distinct edges.

The fuzzy ball atom has no sharp boundaries, but does ha 're definite limits as to the number and energy levels for the electrons residing within each shell.

The fuzzy ball depicts the energy concept of the "electron cloud," which considers the electrons as energy levels around the nucleus. The concept states that the atom is spherical, and the electrons are all over the atom at any one time. The electron cloud concept depicts the relative probable distribution of electrons as they exist at any particular distance from the nucleus at any one particular time. The fuzziness is the image one might see after viewing the atom over an extended period of time, such as a time exposure with a camera.

The number of electrons in each shell is dependent on how far the shell is from the nucleus. And electrons assigned to a specific shell stay in position until forced out by an input of energy. Even more important, a particular shell or energy level does not like to have any electrons missing. If a shell does not have its complete quota of electrons, the atom will "bond" with other atoms by "taking in electrons" or "giving up electrons" or "sharing electrons" to maintain a complete outer shell, thus forming molecules. This is the essence of chemical reactions between and among elements that form compounds.

There are seven possible shells or energy levels for electrons. The lightest atoms have only one shell, which is the innermost shell closest to the nucleus. Other atoms have multiple shells, and the largest and heaviest atoms have all seven shells of electrons. All the electrons in a particular shell have the same energy, and the energy increases for each shell as it is further from the nucleus. The electrons at the greatest distance from the nucleus are the ones with the weakest attraction to the nucleus; thus they are usually the first electrons to be involved in a chemical reaction. These outer electrons then become the **valence** electrons.

In addition, there are subshells for each major shell, known as *suborbitals*. As previously mentioned, **orbital theory** is based on the quantum theory, which states that the positions of electrons cannot be precisely determined in the orbits. However, their positions can be predicted by laws of mathematical probability. Suborbitals take the shape of indefinite spheres and elliptically shaped doughnuts. Each suborbital is assigned a letter, either s, p, d, or f. The orbitals for all the elements are listed at the end of the book. Following is a list of the seven major shells, including the four suborbital energy levels, with the letters assigned for their position from the nucleus. The number of electrons required to fill each shell and suborbital is shown.

Level	Shells	Maximum Number of Electrons	Suborbitals
1	Shell K	2	s2
2	Shell L	8	s2, p6
3	Shell M	18	s2, p6, d10
4	Shell N	32	s2, p6, d10, f14
5	Shell O	32	s2, p6, d10, f14
6	Shell P	18	s2, p6, d10
7	Shell Q	2	s2

Note: In some of the heavier elements, the N and O shells may become the valence shells, even though they are not the outside shells. Some references show the sixth-level "P" shell as having a maximum of 32 electrons, with suborbitals s2, p6, d10, f14.

This is also the *maximum* number of possible electrons in the shells of the known elements.

Up to this point we have been describing single atoms and their electrons. When the electrons of the atoms of two or more different elements are involved, a **chemical reaction** occurs. **Nuclear reactions** involve the particles in the nucleus of the atom, not the electrons. The distinction is fundamental. The former is atomic (or electron) chemistry, while the latter is nuclear chemistry or nuclear physics.

Some confusion exists. For instance, in normal chemical reactions, the electrons of different atoms interact to form molecules of new chemical compounds. As an example, the explosion of gunpowder is a chemical reaction involving the interactions of electrons of different elements, resulting in by-products of energy (the explosion, heat, and radiation) and the gases released. In contrast, the atom bomb or an atomic energy power plant derives its energy from the nucleus, not the electrons. So it is inappropriate to call the "bomb" an atomic bomb, because the electrons of the atom are not the source of the energy. The **fission** (splitting) of nuclei or the **fusion** (combining) of nuclei is the source of the energy that causes the nuclear explosion. A detonation of an "atom" bomb or the production of energy in a nuclear power plant is not a chemical reaction but a nuclear reaction. To the nonscientist there is little distinction between a chemical weapon and a nuclear weapon—both are deadly. But the distinction is fundamental.

THE NUCLEUS AND RADIATION

Many scientists contributed to concepts of **radiation** as produced by the **particles** making up the nuclei of atoms. For example:

While experimenting with cathode ray tubes, Wilhelm Konrad Roentgen (1845–1923) discovered radiation with which he was not familiar. He named these X-rays, using "X" to signify the "unknown."

Antoine Henri Becquerel (1852–1908) discovered **alpha particles**, which are, in essence, the nuclei of the element helium (He^{++} ions). Compared to electrons, they are heavy, only travel short distances, and may be stopped by a sheet of paper.

Ernest Rutherford (1871–1937) is one of the discoverers of **beta particles**, which are high-speed electrons that can travel at the speed of light over great distances. An example is electrons traveling over wires to our homes. High-energy beta particles stripped from the atom can penetrate several sheets of cardboard. An electron has a mass that is almost 2,000 times *less* than the mass of an atom.

Marie Curie (1867–1934) and Pierre Curie (1859–1906) experimented with **gamma rays**, which have a shorter wavelength, travel at the speed of light, and can penetrate several inches of lead, depending on their source and energy level. There are several types of gamma rays. Some examples are the product of a short burst of protons in a **cyclotron** or "atom smasher," cosmic radiation from outer space, and the natural disintegration of the nuclei of radioactive atoms in the Earth. Most of the heavier elements with an atomic number above 88 are **radioactive** to some degree.

We previously mentioned the classic experiment by Ernest Rutherford in the early 1900s which established the concept that most of the mass of an atom exists as a tiny center of positively charged matter, called the nucleus. This nucleus is only about one millionth of the diameter of the whole atom. Rutherford's experiment also established that the negative electrons surrounding the nucleus are at a relatively great distance from this very dense central mass. In addition, the electrons make up only an extremely small fraction of the atom's weight. Thus, most of the atom is empty space, consisting of a tiny, positively charged, relatively heavy, dense core surrounded by distant, and much lighter, negatively charged electrons.

Here is how Rutherford conducted the experiment. He caused high-speed alpha particles (positive helium nuclei) to bombard a thin sheet of gold foil that was only one fifty-thousandth (1/50,000) of an inch thick. Since the thickness of this thin foil was only about two thousand atoms of gold, it should have allowed all the alpha particles to pass through and be detectable on the other side of the foil. As it turned out, 1 out of every 10,000 alpha particles, as they passed through the foil, were deflected sideways away from the center of the target located behind the foil. Also, 1 out of every 20,000 particles bounced back from the foil toward the source of the alpha particles, and thus did not pass through the gold foil at all. This indicated that there was some small, very heavy mass in the gold atoms that caused the deflection of the alpha particles. Since the vast majority of the alpha particles passed straight through the foil to hit the center of the target, the atoms of gold must be composed of mostly empty space. This led to our current conception of the structure of the atom.

An interesting discovery by Rutherford and other scientists is that there is, comparatively, a great distance between the nucleus and the electrons in atoms. One way to understand this great distance is to imagine a baseball on the floor of your living room as the nucleus of an atom. The surrounding electrons are about the size of peas and are orbiting in three dimensions at about a 10-mile radius from the baseball. This means that an atom with a 20-mile diameter would have a nucleus about the size of a baseball. Yes, atoms, molecules, and all matter on the Earth, including you, are mostly empty space.

Many other experiments verified the concept that the nucleus consists of two major types of particles: **protons**, which carry a positive charge, and **neutrons**, which have a similar mass to the protons, but have no electrical charge. Thus, the total mass of a nucleus consists of the total number of both protons and neutrons. Together, they make up all but a tiny fraction of the weight of an atom. The protons, by themselves, are the source of the positive electrical charge of the nucleus, which is balanced by the negatively charged electrons.

The science of particle physics continues to study electrons, protons, and neutrons, which are considered **subatomic** particles. The quest continues for even smaller subatomic, or rather subnuclear, particles. Most subnuclear particles are fleeting in time of existence, are practically weightless, and are thus very difficult to detect and measure.

The search continues. In the February 11, 1996, issue of the *Houston Chronicle* there is an article by Faye Flam, of the *Philadelphia Inquirer*, titled "Physicists on Trail of Particles More Minuscule than Quarks." The article describes how scientists are seeking particles-within-particles using the 4.5-mile-long Tevatron at the Fermi National Laboratory near Chicago. The Tevatron is one of many tools, such as the atom smasher, used by high-energy physicists. The Tevatron is a long tunnel through which scientists send high-speed protons, while from the other direction they send high-energy **antiprotons** (antimatter). When the two meet, tremendous bursts of energy and smaller particles result. Both protons and antiprotons are made of **quarks**. When their quarks collide, there is evidence of smaller particles. Quarks are hypothetical entities that carry very small electrical charges and are considered the major bit of the smallest bits of matter. They serve the purpose of facilitating calculations. Many smaller, more fleeting bits of matter than the quark exhibit properties of both particles and waves. There are many such wave/particles with odd names that have been detected by more elaborate experimental equipment. By separating and measuring smaller and smaller bits of matter, we develop new understanding of the origin and nature of matter.

CHEMICAL BONDING

As we have stated, chemical reactions are the interactions of electrons that "join" the shells surrounding the atoms of different elements. How these electrons of different atoms combine to form molecules is the essence of the science of chemistry.

Let's define the difference between an atom and an ion. The atom has a neutral electrical charge because of the equal number of electrons and protons. The ion is an atom that has either a negative or a positive charge due to the gain or loss of one or more electrons.

Atoms have the unique capacity to maintain outer shells, each having a complete number of the required electrons (2, 8, 18, or 32). For instance, if an atom of an element has a single electron in its outermost shell, it will reactively give that electron up or share it with another atom that requires an electron to complete its shell. A simple example: the element sodium is a very reactive metal with one electron in its outer shell which it wants to give up so it can have a complete shell at the next lower level. It will react with an element, such as chlorine, that has only seven electrons in its outer shell. Chlorine "wants" to gain one electron to complete its outer shell at eight. Thus, both elements react to complete their outer shells. Sodium gives up an electron, and chlorine accepts it, creating a new molecule of sodium chloride. When this occurs, the atoms have now become ions (individually they have a charge) and the joining between the two is an **ionic bond**. The ionic bond holds the atoms together to form the new molecule of a compound called sodium chloride or table salt: (Na + Cl = NaCl). Table salt has none of the characteristics of the two original elements.

There is another type of bond called a **covalent bond**, in which two or more atoms share the electrons rather than give up or receive electrons as in ionic bonding. The most common element that forms covalent bonding is carbon, which has four electrons in its outermost shell, thus permitting it to mutually exchange electrons with other elements. The molecules and compounds formed by covalent reactions involving carbon are mostly organic compounds. Organic chemistry is the study of carbon molecules and compounds that compose living organisms, i.e., plants, animals, wool, food, and so forth, and all the hydrocarbon compounds we use, such as oil, gasoline, grease, and cosmetics. Hydrogen and carbon can form many interesting and useful compounds.

The type of chemical bonding determines the physical properties of materials on earth. For example, the less rigid and firm the bonds are between atoms and molecules, the more likely the material will be soft and flexible. The more firm or tight the bonds, or compactness of the atoms, the more solid the material will be. The types of bonding, as well as density, determine if a substance will be a gas, liquid, or solid.

This book deals with aspects of **inorganic chemistry** rather than **organic chemistry**. It presents the physical properties, chemical characteristics, atomic structures, sources and abundance, history, uses, and hazards of the Earth's elements. The classifications used in this book are representative of the structure of atoms of each specific element, not the structure of molecules. Although the study of bonding and molecular structure of elements and compounds is important, this book will emphasize the elements and their atoms.

3

THE PERIODIC TABLE OF THE CHEMICAL ELEMENTS

HISTORY

Without a doubt, the Periodic Table of the Chemical Elements is the most elegant organizational chart ever devised. Conceptually, many individuals recognized that certain chemical elements have similar characteristics to other chemical elements. This was accepted even though the atoms of a particular element are different from atoms of related elements. How do we determine what the relationships of these similar characteristics are when compared with different elements?

In 1829 Wolfgang Dobereiner (1780–1849) recognized that similar elements occurred in groups of three, which he called triads. Julius Lothar Meyer (1830–1895) and John Alexander Reina Newlands (1837–1898) both developed charts based on the atomic weights of the elements as known in their time. They recognized the "octave" (eight) nature of repeating characteristics of several elements, which preceded the concept of Periods.

By all standards, the father of the modern Periodic Table was a Russian chemist, Dmitri Ivanovich Mendeleyev (1834–1907). In 1871 Mendeleyev arranged the elements not only by their atomic weights in horizontal rows, or Periods, but also in vertical columns, or Groups, by their **valences** and other characteristics.

Remember, valence is the combining power, in whole numbers, of one element with another. The most common combining ratio of atoms of different elements to form molecules is 1:1, but other ratios are possible. For instance, hydrogen (H) and chlorine (Cl) both have a valence of 1, so they combine to form a molecule of HCl, hydrogen chloride or hydrochloric acid. Nitrogen has a combining power (valence) of 3. When hydrogen combines with nitrogen, it takes 3 H atoms for each N atom to form the resulting compound, ammonia (NH_3).

Mendeleyev's chart had vacant spaces that provided him and other scientists a plan to predict where new elements would fit as new discoveries were made. As the chart began to fill in, some problems became evident. The British physicist Henry Gwyn Mosely (1887–1915), by using X-rays, identified the increasing positive charges of nuclei as the atomic weights of atoms increased. His discovery led to a correction of the Periodic Table. Instead of arranging the elements according to their atomic weight, he arranged them by their atomic number, i.e., number of positive protons in the nucleus. And, this is how it exists today.

Let's take a more detailed look at the Periodic Table and discover its great symmetry and usefulness.

RULES FOR CATALOGING THE ELEMENTS

1. The order in which the elements appear in the Periodic Table follows specific rules.

2. Changes in properties of the elements repeat in very orderly ways.

3. The arrangement of the elements in the Table is according to:

a) The electron configuration of the atoms of the elements. The electrons are the negatively charged particles in shells surrounding the nucleus. The arrangement of the electrons in the outermost shells determines the element's "combining power" or valence, and;

b) The number of positively charged particles (protons) in the nucleus of an atom determines that atom's atomic number. In other words, the number of protons determines the atomic number, which is also referred to as the proton number.

c) The total number of positive protons plus the total number of neutrons, which have no electrical charge, are found in the nucleus. Together, the total number of both the protons and neutrons determines the atomic weight (atomic mass) of each atom. For example, at. wt. – number of neutrons = at. no.

4. Periods are the rows that run left to right.

5. Both the atomic numbers (protons) and electrons increase in number from left to right in each Period.

6. When a principal energy level (shell) receives its full complement of electrons, e.g., inert noble gases in Group 18 (VIIIA), a new row begins, which is the start of a new Period.

7. The vertical columns represent families of elements that exhibit similar characteristics. These families of elements are called Groups. Elements in the same Group, in general, exhibit similar combining powers (valence), but do not exhibit the same degree of reactivity to other elements.

8. Most elements in the same Group (family) have the same number of valence electrons in their outermost shell.

9. In general, for Groups, when reading the Table from left to right, the number of valence electrons in the outer shell of elements increases. For instance, lithium is located at the start of Period 2 and has 1 electron in its outer shell and is in Group 1 (1A), and neon, located at the end of Period 2 in Group 18 (VIIIA), has 8 electrons in its completed outer shell.

There are several versions for numbering the Groups from left to right. The International Union of Pure and Applied Chemistry (IUPAC) uses a new notation system of arabic numerals, 1 to 18 for the Groups. Roman numerals are used in the old IUPAC form. Hydrogen is located on the far upper left of the Table in the first Period in Group 1. Compare this placement with neon, which has eight electrons in its full outer shell. As an inert gas, neon is in Group 18 (VIIIA) on the far right column of the table.

10. **Metals** (on the left side of the Table), in general, have fewer electrons in their outer valence shell than do **nonmetals**.

11. Metals give up valence electrons or share electrons with nonmetals. The most active metals are the ones on the left side of the Table that have the least number of valence electrons.

12. Most nonmetals located in the Groups on the right side of the table have more electrons in their outer valence shells than do metals.

13. The most reactive nonmetals are those in Group 17 (VIIA) on the right side of the Table (see exception below). They tend to accept valence electrons from the metals to complete their outer valence shells from 7 electrons to full outer shells of 8 electrons.

14. The atomic and ionic sizes of elements, in general, increase from the top of the Table to the bottom as the atomic mass and the number of electron shells increase.

15. Thus for both metals and nonmetals the atomic size increases as the atomic mass (weight) and atomic number increase.

16. The neutron is a fundamental particle of matter, found in the nucleus. The neutron has about the same mass as the proton, but unlike the proton the neutron has no electrical charge.

17. The total *atomic mass* (weight) of an atom consists almost entirely of the total mass of both the protons and the neutrons. Electrons (negatively charged particles) have a mass of less than 1/2000 that of the proton (actually 1/1837). Therefore, the electron's mass is negligible when considering the total atomic mass (weight) of an atom.

18. The *atomic number* is the number of protons (positive charges) in the nucleus.

19. Exceptions:

a) Elements in the far right column, Group 18 (VIIIA), all have completed outer shells of 8 valence electrons. Thus, they do not easily react with other elements. They are known as the noble gases or inert elements.

b) Two elements that do not fit the primary pattern of the Table are hydrogen and helium.

Hydrogen, although not a metal, is usually placed in the Table to head the first group, the alkali metals. Hydrogen has one electron in the K shell, which is its first and only shell. With one valence electron, hydrogen usually reacts like a metal because it tends to "give up" or share its single electron to become a positive ion.

On the other hand, hydrogen, with its single electron in the K shell, could also be placed above the halogens in Group 17 (VIIA), which indicates it needs an electron to complete its outer valence orbit. Hydrogen can collect an electron to form a **hydride** with a metal, e.g., $2Na + H_2 \rightarrow 2NaH$ (sodium hydride). So hydrogen can either give, share, or receive an electron.

Helium is placed at the top of the far right column or group 18 (VIIIA) because it has 2 electrons in the K shell (the first shell), which completes its outer valence shell. Helium is inactive; therefore it is included in the Group with the noble inert gases, even though its completed outer shell has 2 instead of 8 electrons.

20. The **transition elements** are found in three series in the center of the Periodic Table starting at Group 3. The first series is in Period 4, from scandium (Sc^{21}) to zinc (Zn^{30}). The second series is in Period 5, yttrium (Y^{39}) to cadmium (Cd^{48}). The third series is a bit different. It is in Period 6 following the **lanthanide series**, and starts at Group 4 at the element hafnium (Hf^{72}) and continues to include mercury (Hg^{80}). The transition elements represent a change in the chemical characteristics of the elements in these series from strong metals to nonmetals. They show a gradual shift from being strongly **electropositive**, i.e., giving up valence electrons, as do elements in Groups 1 (IA) and 2 (IIA). The shift (transition) continues to the strongly **electronegative** elements, i.e., gaining or sharing valence electrons, as in Groups 15, 16, and 17. Thus, they progress from having some properties and characteristics similar to those of metals to several properties more like those of **metalloids (semiconducting** elements), and nonmetals.

21. Two other series of elements that appear separately in the Periodic Table are the **lanthanide series** and **actinide series**. These series of elements do not fit the normal periodic order of the Periodic Table.

   a) The lanthanide series are also considered "metal-like" because they have two electrons in their outer shells. Because they are difficult to find, they were considered scarce. However, they are not scarce, but they are still called **rare-Earth elements**. There are fifteen elements in this series starting with Group 3 in Period 6. They include the elements lanthanide (La^{57}) to lutetium (Lu^{71}).

   b) The actinide series are both "metal-like" and radioactive. This Series also starts at Group 3, but in Period 7. It includes the element actinium (Ac^{89}) and ends with ununbiium (Uub^{112}). They are unstable and radioactive.

22. The transuranic elements are those with an atomic number higher than uranium (U^{92}). They include the actinides above uranium. Most are man-made, very unstable, and radioactive with very short half-lives. The actinides with higher atomic masses are so unstable that they exist only for microseconds or minutes, and only a few atoms of some have been artificially created.

ISOTOPES

An **isotope** is a different form of an element, similar to a different species of plant or animal. An isotope of an element has the same atomic number as the regular atom of that element, which is the number of protons in the nucleus, but has a different number of neutrons and thus a different atomic weight. An isotope of an element has a different atomic mass, but maintains basically the same chemical characteristics. An isotope of an element occupies the same position in the Periodic Table as does the most common form of that element. Not all elements have isotopes. In these cases, the atoms of these elements all have the same number of neutrons and atomic weights.

The most common form of an element that makes up the majority of the atoms of that element is considered representative of that element. And, the atomic weight of the most represented form of that element is identified by parentheses around the number in the Periodic Table. All other forms of that element that exist (with different atomic masses) and make up only a small fraction of the total atoms of that particular element are called isotopes.

For most elements the majority of its atoms are of the main type as listed in the Periodic Table. Isotopes make up only a very small fraction of the total number of atoms for a particular element. Some examples of isotopes are: hydrogen-2 (called deuterium or heavy hydrogen), carbon-14 (which is a radioactive tracer), and uranium-235 (a fissionable isotope of uranium-238). These isotopes, with atomic masses different from the major type of atoms for these elements, compose a very small fraction of the total atoms that make up these elements.

There are three types of isotopes: (1) naturally stable and nonradioactive, (2) naturally radioactive, and (3) artificially radioactive. Artificially radioactive isotopes are products of nuclear bombardments in **nuclear reactors**, nuclear particle accelerators (atom smashers), nuclear bombs, and similar sources of high-energy particles.

Eighty-three of the most abundant elements have one or more isotopes composed of atoms with different atomic weights. Twenty-one elements have no isotopes; i.e., their atoms all have the same, common atomic weight (mass).

Isotopes of the chemical elements found on Earth are known to exist in space, on the stars, and on other planets. The ratio or proportion of these extraterrestrial isotopes to their main atoms is different than it is on Earth.

RADIOACTIVITY

Atomic or nuclear emission, which is **alpha** (α), **beta** (β), or **gamma** (γ) radiation, results from both natural and artificial or man-made sources. The radiation results from a decay process of the nuclei. For example, the decay of radium-266 emits natural alpha particles (He^{++}, helium nuclei), and a new product, radon-222, is the result. Radon-222, in turn, decays to other products until the stable isotope of the element lead-206 and the resultant energy is a final product.

The time it takes for this radioactive decay to occur is the **half-life** for that isotope. The half-life is the time required for the activity to decrease to one-half of the radioactivity of the original isotope, followed by a decrease of one-half of that remaining half, and then half of that remaining half of the nuclei, etc. This means that the nuclei of radioisotopes do not decay all at once, but rather in random intervals. The half-life is the time required for half of the nuclei to undergo decay to more stable forms of atoms—giving off radiation in the sequences of decay. Interestingly, scientists cannot measure or predict the half-life of any particular single nucleus. We do not have the computing power for this analysis, so we work with averages.

The half-lives of some radioisotopes are measured in billions of years; for others, the half-life is measured in fractions of seconds. Following are some examples of the half-lives of a few isotopes: uranium-238 = 4.6 billion years; carbon-14 = 5730 years; strontium-90 = 38 years; phosphorus-32 = 14.3 days; radon-222 = 3.8 days; uranium-239 = 23.5 minutes.

Radiation is one area of science not well understood by the lay public, and often the media information relating to radioactivity is misleading and misunderstood. To some extent the topic of radioactivity and radiation has become a political issue. The general public is somewhat "scientifically illiterate" about radiation. It does not have a very clear understanding of the physical nature, sources, uses, benefits, and dangers of radiation and radioactivity. We can all learn more about radioactivity so it can be used for the benefit of mankind without undue fear. After all, it is a very natural process that is universal. It takes place both inside and on the surface of our Earth. It not only exists in space, but it is penetrating our bodies at all times. It is part of all life.

Some of the sources of radiation that affect us are the following:

1. Potassium is essential in our diets and is found in many foods. Our bodies cannot distinguish between potassium-39 (nonradioactive) and radioactive potassium-40 in our foods. The radioactive potassium-40 makes up almost one-fourth of all the atomic radiation we receive.

2. Radon gas that seeps into our homes, schools, and offices is produced by the natural decay of uranium in the ground. Radon gas is thought to be a cause of some cancers—particularly lung cancer. Kits are available for testing the levels of radon that might exist in your home.

3. There are other sources of radiation from the decay of radioactive elements in the Earth's crust.

4. We receive radiation from outer space as cosmic rays, solar radiation, and upper atmosphere radiation. The higher the altitude at which you live, the greater your exposure to cosmic radiation from space. Since nuclear radiation accumulates in our bodies over time, people living in Denver, Colorado, receive more radiation in their lifetimes than do people living in areas at sea level.

5. In the past we received some radiation from the testing and explosion of nuclear weapons. Strontium-90 was of particular concern because it has a long half-life and becomes concentrated in the food chain, particularly in milk. The ban on atmospheric testing of nuclear weapons has reduced this hazard.

The greatest hazards from radiation are the changes caused in the cells of the body, including the reproductive sperm and egg cells. Radioactive potassium is thought to be a main source of mutations (genetic changes) in plant and animal cells. This may have some relationship to the evolution of plant and animal species since all plants and animals require potassium to survive. As mentioned earlier, living organisms cannot tell stable potassium from the radioactive form in their diets.

Our exposure to man-made radioactive sources, such as from nuclear power plants, is negligible when compared to the total radiation we receive. Man-made radiation accounts for less then 3% of the total radiation we receive in the United States, but in some countries the figure is higher. *The vast majority of the 3% of the MAN-MADE doses of radiation we receive in our lifetime results from medical uses, while the vast majority of the 97% of the TOTAL exposure to ALL radiation we receive comes from natural sources.*

In summary: The structure and recurring characteristics of elements are represented in the "catalog" we refer to as the Periodic Table of Chemical Elements. To review:

1. The negatively charged **electrons** surround the positively charged nucleus. Electrons can exist in shells or energy levels orbiting around the nucleus, or they can easily be stripped from the atom and exist in a free state as electricity. An electron is 1/1837 the weight of a proton. Even so, an electron's electrical charge balances the positive charge of a proton. Also, the electrons are about 1 million times distant from the nucleus of an atom as compared to the diameter of the nucleus. In the neutral atom, the number of electrons equals the number of protons.

2. The positively charged **protons** are compacted in a tiny, dense center of the atom called the nucleus. The number of protons in the nucleus determines the atomic number for each element, but not its mass. The Periodic Table lists the number of protons in progression from the first number, hydrogen (H^1), to the most recently discovered element with the highest atomic number, unununium (Uuu^{111}).

3. The **neutron** was more difficult to identify because it had no electrical charge. Neutrons are found in the nucleus with the protons. They have approximately the same mass as the protons, and together they make up

the atomic mass of the atoms for each of the elements. To determine the number of neutrons in the atoms of an element, one can subtract the atomic number (protons) from the total atomic mass (atomic mass – number of protons = number of neutrons).

4. Ions are atoms that have gained or lost electron(s) in a chemical reaction to form a new substance or compound. Thus, one might think of ions as atoms with electrical charges.

5. Isotopes are atoms with more than the usual number of neutrons in their nuclei. Their atomic number does not change, but their atomic mass does. Many isotopes of atoms with high atomic masses are radioactive.

The background presented here will aid you in learning how to use the Periodic Table of the Chemical Elements. Once you study and understand the organization of this remarkable chart, you can glean a great deal of information about the Earth's chemical elements. The structure and characteristics of elements may be identified by their position in the Table. How a given element will react with other elements can be predicted by its placement in relation to other elements in the Table. The Periodic Table contains much intriguing and useful information for you to discover.

The Periodic Table of the Chemical Elements is the "Rosetta Stone" that decodes the nature of chemistry.

4

ALKALI METALS AND ALKALI EARTH METALS

Metals are elements that form positive ions when combining with non-metals. Most metals have less than four electrons in their outer valence shell. Some metals are unique since they can also use electrons in the next to the outer shell for forming bonds. Metals are classified as *electropositive* since they "give up" or share electrons in chemical reactions, becoming positive ions. Or if in an electrolytic solution, they are called **cations** because they collect at the negative pole or cathode. Cations have more protons in their nuclei than electrons in their shells. Also, their electrons are more strongly attracted to the positive nucleus; thus metallic ions are smaller than atoms of the same elements.

Anions are just the opposite. They are negative ions that have more electrons than protons; thus they have a negative charge and collect at the positive anode. For anions the electron attraction to the nucleus is weaker; thus, in general, they are larger than atoms of the same elements.

As discussed in Chapter 3, the metals are grouped on the left side of the Periodic Table. As you move to the right side of the Table, the very reactive metals change to less active metals, or **transition metals**, and then to **metalloids**, **semiconducting** elements, and then to **nonmetals**, whose outer valence shells tend to gain electrons when reacting with metals. As the number of electrons in the outer shells increases, the distance between the nucleus and the electrons also increases. The further the electrons are from the nucleus, the weaker the attraction between the electrons and the positive nucleus. Thus, the ionization energy increases from left to right in the Table.

Approximately 75% of all the elements on the Earth are metals. They are **crystalline** solids, which range from hard to butter-like soft, conduct electricity, and have a metallic luster when cut. Those located at the far left have only one electron in their outer shell, are very reactive, and are not usually found in pure form. Instead, they are found in compounds, **minerals** or **ores** that must go through processes that extract the pure metal from the other substances.

Metals may be classified by a variety of categories. Some metals fit more than one of the following categories:

> **Alkali Metals**—Group 1 (IA), soft, silvery, and very active. They are all solids and have 1 electron in their outer valence shell. The alkali metals begin each period, 2 to 7, on the Periodic Table. Even though hydrogen may be considered a nonmetal, it is included in Group 1 with the alkali metals because it has a single electron in its valence shell and, thus, may act like a metal.
>
> **Alkali Earth Metals**—Group 2 (IIA), shades of white to colors, **malleable**, **machinable**, less active than alkali metals. They are all solids and have 2 electrons in their outer valence shell.
>
> **Transition Metals**—found in the groups located in the center of the Periodic Table, plus the **lanthanide** and **actinide** series. They are all solids, except mercury, and are the only elements to use shells other than their outer shells for giving up or sharing electrons.
>
> **Noble Metals**—refers to several unreactive metals that do not easily dissolve in **acids** or **oxidize** in air, e.g., platinum, gold, and mercury. They include the platinum group of metals (see next). They are called "noble" because of their resistance to reacting with other elements.
>
> **Platinum Metals**—include unreactive transition elements located in Groups 8, 9, and 10 of Periods 5 and 6. They have similar chemical properties. They are ruthenium (Ru^{44}), rhodium (Rh^{45}), palladium (Pd^{46}), osmium (Os^{76}), iridium (Ir^{77}), and platinum (Pt^{78}).
>
> **Rare Metals**—a loose term for less well known metallic elements, including the alkaline-Earth metals and so-called **rare Earths**. The rare Earths aren't really scarce; they were just difficult to isolate and identify.
>
> **Lanthanide Metals**—rare Earths, including elements of atomic numbers 57 through 71.
>
> **Actinide Metals**—includes elements with atomic numbers from 89 to 111. Also includes the transuranic elements (beyond uranium [U^{92}] to Lr^{103}) and

the superactinides (numbers 104 to 112), which are unstable, radioactive and man-made.

Light Metals—a general term for elements that are light but strong enough for construction, e.g., aluminum, magnesium, titanium, and beryllium.

Heavy Metals—a general term for metals with atomic weights greater than 23.

Metals are extremely important not only for chemical reactions but also for the health and welfare of plants and animals. Some metals, e.g., mercury, lead, cadmium, barium, beryllium, radium, and uranium, are very toxic. Some examples of metals required for good nutrition, even in trace amounts, are, iron, copper, cobalt, potassium, sodium, and zinc. Some metals at the atomic and ionic level are crucial for the oxidation process that **metabolizes** carbohydrates for all living cells.

Group 1 Elements
THE ALKALI METALS

The alkali metals are the elements in Group 1 (IA), from Periods 2 through 7 (rows). Hydrogen (H^1) is not really an alkali metal, but it best fits Group 1 since it gives up its single electron found in its valence (outer) shell to form a positive ion, as do the other alkali metals. Thus, it is assigned to Group 1 in Period 1.

The alkali metals are electropositive and form cations (positively charged ions) by either giving up or sharing their single electron located in their outermost electron shell. The other elements of Group 1 are: lithium (Li^3), sodium (Na^{11}), potassium (K^{19}), rubidium (Rb^{37}), cesium (Cs^{55}), and francium (Fr^{87}). Following are some characteristics of the Group 1 alkali metals:

1. With the exception of francium, they are all relatively soft.

2. They are silvery in color, especially after a fresh cut.

3. Their melting and boiling points become lower as their atomic weights increase; i.e., they proceed down Group 1, from Period 2 to Period 7. For example, lithium melts at about 180 degrees Celsius while francium melts at an stimated temperature of about 27 degrees Celsius, which is just above room temperature.

4. Their atomic volumes become larger as their weights increase.

5. They produce distinctive colored flames when burned: lithium = crimson; sodium = yellow; potassium = violet; rubidium = purple; cesium = blue; the color of francium's flame is not known. Many of francium's characteristics have not been determined owing to its short **half-life**.

PERIODIC TABLE OF THE ELEMENTS

GROUPS / PERIODS	1 IA	2 IIA	3 IIIB	4 IVB	5 VB	6 VIB	7 VIIB	8	9 VIII	10	11 IB	12 IIB	13 IIIA	14 IVA	15 VA	16 VIA	17 VIIA	18 VIIIA
1	1 **H** 1.0079																	2 **He** 4.00260
2	3 **Li** 6.941	4 **Be** 9.01218											5 **B** 10.81	6 **C** 12.011	7 **N** 14.0067	8 **O** 15.9994	9 **F** 18.9984	10 **Ne** 20.179
3	11 **Na** 22.9898	12 **Mg** 24.305											13 **Al** 26.9815	14 **Si** 28.0855	15 **P** 30.9738	16 **S** 32.066(6)	17 **Cl** 35.453	18 **Ar** 39.948
4	19 **K** 39.0983	20 **Ca** 40.08	21 **Sc** 44.9559	22 **Ti** 47.88	23 **V** 50.9415	24 **Cr** 51.996	25 **Mn** 54.9380	26 **Fe** 55.847	27 **Co** 58.9332	28 **Ni** 58.69	29 **Cu** 63.546	30 **Zn** 65.39	31 **Ga** 69.72	32 **Ge** 72.59	33 **As** 74.9216	34 **Se** 78.96	35 **Br** 79.904	36 **Kr** 83.80
5	37 **Rb** 85.4678	38 **Sr** 87.62	39 **Y** 88.9059	40 **Zr** 91.224	41 **Nb** 92.9064	42 **Mo** 95.94	43 **Tc** (98)	44 **Ru** 101.07	45 **Rh** 102.906	46 **Pd** 106.42	47 **Ag** 107.868	48 **Cd** 112.41	49 **In** 114.82	50 **Sn** 118.71	51 **Sb** 121.75	52 **Te** 127.60	53 **I** 126.905	54 **Xe** 131.29
6	55 **Cs** 132.905	56 **Ba** 137.33	57 **La** 138.906 ★	72 **Hf** 178.49	73 **Ta** 180.948	74 **W** 183.85	75 **Re** 186.207	76 **Os** 190.2	77 **Ir** 192.22	78 **Pt** 195.08	79 **Au** 196.967	80 **Hg** 200.59	81 **Tl** 204.383	82 **Pb** 207.2	83 **Bi** 208.980	84 **Po** (209)	85 **At** (210)	86 **Rn** (222)
7	87 **Fr** (223)	88 **Ra** 226.025	89 **Ac** 227.028 ▲	104 **Und** (261)	105 **Unp** (262)	106 **Unh** (263)	107 **Uns** (264)	108 **Uno** (265)	109 **Une** (266)	110 **Uun** (267)	111 **Uuu** (272)	112 **Uub**	113 **Uut**	114 **Uuq**	115 **Uup**	116 **Uuh**	117 **Uus**	118 **Uuo**

TRANSITION ELEMENTS

6 ★ Lanthanide Series (RARE EARTH)

58 **Ce** 140.12	59 **Pr** 140.908	60 **Nd** 144.24	61 **Pm** (145)	62 **Sm** 150.36	63 **Eu** 151.96	64 **Gd** 157.25	65 **Tb** 158.925	66 **Dy** 162.50	67 **Ho** 164.930	68 **Er** 167.26	69 **Tm** 168.934	70 **Yb** 173.04	71 **Lu** 174.967

7 ▲ Actinide Series (RARE EARTH)

90 **Th** 232.038	91 **Pa** 231.036	92 **U** 238.029	93 **Np** 237.048	94 **Pu** (244)	95 **Am** (243)	96 **Cm** (247)	97 **Bk** (247)	98 **Cf** (251)	99 **Es** (252)	100 **Fm** (257)	101 **Md** (258)	102 **No** (259)	103 **Lr** (260)

6. They react with water and weak acids, some explosively.

7. Their alkalinity is above *p*H 7 (7 is neutral in the *p*H acid/base scale). Their *p*H increases above 7 as they proceed down Group 1 and as their atomic weights become greater, with the exception of francium.

Following is detailed information about the seven alkali metals, including hydrogen:

HYDROGEN (Alkali Metals)
SYMBOL: H   ATOMIC NUMBER: 1   PERIOD: 1

COMMON VALENCE: 1   **ATOMIC WEIGHT:** 1.008   **NATURAL STATE**: Gas
COMMON ISOTOPES: ^{2}D deuterium (H-2) (1 proton + 1 neutron, rare, found in "heavy" water), and ^{3}T tritium (H-3) (1 proton + 2 neutrons, man-made by nuclear reactions and radioactive). **PROPERTIES:** Hydrogen's molecular formula as a diatomic gas (composed of two atoms) is H_2. Density: 0.08999 (air = 1.0); freezing point: –259°C; boiling point: –252°C; (**absolute zero** = –273.13°C or –459.4°F).

CHARACTERISTICS

The atom of the most abundant form of hydrogen consists of a nucleus with 1 proton and 1 electron located in the first (K) shell.

H_2 is a diatomic gas molecule that is odorless, tasteless, and colorless. Even though it is included in Group 1 with the alkali metals, hydrogen is a nonmetal.

H_2 is slightly soluble in water, alcohol, and ether. It is noncorrosive, but can permeate solids better than air. Hydrogen has excellent **aDsorption** capabilities where it is attached and held to the surface of some substances. (Note: Not the same as **aBsorption**, where one substance intersperses another.)

ABUNDANCE AND SOURCE

Hydrogen is the tenth most abundant element on the Earth, and it is the most

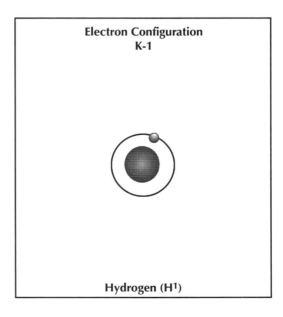

Electron Configuration
K-1

Hydrogen (H^1)

abundant element in the universe. It occurs mainly in a combined form as water on the Earth. Hydrogen is also found in organic and hydrocarbon compounds. Only traces of hydrogen are found in the atmosphere. Since hydrogen is the lightest gas, it escapes from the Earth's gravity when in the atmosphere.

Hydrogen may have been formed by the "big bang," accounting for over 98% of all the atoms in the universe at that time. Today hydrogen makes up over 90% of all the atoms and about 75% of the entire weight (mass) of the universe. The big bang is one theory for the origin of the universe. As the theory states, it all started many billions of years ago when an incredibly small, but dense, ball of matter exploded and dispersed matter in an ever-expanding pattern, which continues to expand even today.

Hydrogen produced by the big bang is possibly the original source, or building block, for all other elements in the universe, including Earth.

HISTORY

Hydrogen was first produced by Theophrastus Bombastus von Hohenheim (known as the alchemist Paracelsus) (1493–1541) by mixing metals with acids. He did not know that the explosive gas produced by this chemical reaction was hydrogen.

In 1766, Henry Cavendish (1731–1810) identified the gas from this reaction as "fire air" and determined that the gas, when burned in air, produced water. He is given credit for the discovery of hydrogen. In 1783 Antoine-Laurent Lavoisier (1743–1794) gave the gas its name, *hydrogen*, which means "maker of water."

Hydrogen gas can be produced by gasification of coal by the addition of hydrogen to the coal to form methane (CH_4), which is then used as a fuel or chemical (**hydrogasification**). Another method is a reaction of steam on hot coal in the presence of air (the synthesis of gas). This is how the "old gasworks" produced cooking gas, which was called "water gas" before natural gas became widely used for heating and cooking.

Natural gas is a low-molecular-weight hydrocarbon consisting of about 85% methane (CH_4), which is recovered as an underground gas in areas where petroleum is found. It is also economically produced by conversion of crude oil to gas.

The most efficient production method for hydrogen (but not the most economical) is **electrolysis** of water, H_2O, where the hydrogen gas is separated from the oxygen gas by passing an electrical charge through the water. Other methods of production are being explored.

COMMON USES

Hydrogen is an excellent and very reactive **reducing agent** used in many industries.

Production of Ammonia: Ammonia (HN_3) is produced by the **Haber process** and is the fifth most produced industrial compound. Ammonia has many uses, including in fertilizer, explosives, and rocket fuel. It is explosive when mixed with silver or mercury and as ammonium nitrate.

Ethanol (Ethyl Alcohol Made from Grains): Ethanol (C_2H_5OH) is one of many types of alcohol. Grain alcohol can be produced by fermentation of agricultural waste or corn. Another method is the hydration of ethylene, i.e., the reaction of water on ethylene (C_2H_4) where the alcohol molecule is formed; the water is then split off by heat.

One hundred percent pure ethyl alcohol is a colorless, volatile liquid with a pungent taste. One hundred proof alcoholic drinks are just 50% ethanol. When consumed, ethanol is a depressant and is habit-forming. It rapidly oxidizes in the body, but even small amounts cause dizziness, nausea, headaches, and loss of motor control.

Proof was how whiskey salesmen of the Old West demonstrated that their product was potent. They would place some gun powder in a tin dish, then pour on some of their whiskey, and if a match would ignite the mixture it was "100% proof" of its strength. It just happens that 100-proof whiskey is 50% ethanol.

Hydrogenation of Vegetable Oils: Hydrogenation is the process by which:

1. Hydrogen is added to the carbon double bonds of **unsaturated** or **poly-unsaturated** molecules where not all these double carbon bonds are used—and thus they react with other elements, e.g., the "good" vegetable liquid fats and fish/chicken types of fats.

2. The added hydrogen forms saturated molecules where all single-carbon valence bonds become satisfied or attached to other atoms, thus producing "bad" fats, such as solid animal and dairy fats that may increase cholesterol and cause atherosclerosis (hardening of the arteries).

Hydrogenation "saturates" or takes up the double bonds of unsaturated liquid vegetable oils to form solid (saturated) fats, which may be more appealing and have a longer shelf life, but are not necessarily a better dietary choice. An example is converting liquid vegetable oils to products such as margarine. There is some recent concern that "trans fats" in processed foods may boost the bad cholesterol level in the blood and contribute to heart problems. There is some evidence that trans fatty acids are just as harmful as hydrogenated oils or shortening. Non-tropical vegetable oils such as canola oil, peanut oil, and corn oil seem to deposit the least cholesterol in human arteries.

Hydrocracking of Petroleum and Coal: Hydrogen is used for a process called destructive hydrogenation that breaks down large hydrocarbon molecules to form more useful liquid and gaseous fuels.

Reducing Agent: Hydrogen provides electrons to compound molecules for a desired chemical reaction. Reduction is the opposite of oxidation (where oxygen accepts electrons in a chemical reaction).

Lighter-than-Air Balloons, such as Weather and Sonar Balloons: Hydrogen is no longer used for passenger balloons (see Hazards below).

As a Rocket Fuel: Large amounts of liquid hydrogen, along with liquid oxygen, are used in the U.S. Space Program.

As an "Ion" Fuel for Nuclear Rocket Engines: By stripping the single electron from a hydrogen atom, you end up with a light positive particle (proton) that can be ejected at great velocity from the rocket engine.

As a Possible Fuel for "Fuel Cells": Hydrogen and oxygen gases can be combined in a **fuel cell** to produce electricity with heat and water as the by-products. This reaction is the opposite of the electrolysis of water where electricity separates the water molecules to form H_2 and O_2. On May 14, 1996, the Daimler-Benz German automobile maker, in cooperation with the Ballard Power Systems Co. of Canada, announced development of the first fuel cell–powered passenger car. A similar fuel cell power system developed by the Canadian company is being tested on several buses in cities in the United States. The source of the fuels, at least for the near future, will be natural gas for the hydrogen and air for the oxygen. The fuel cell power system produces very little pollution. The fuel cell is a reliable source of both electricity and drinking water for astronauts. However, both the U.S. government and energy companies have reduced their research support for fuel cells designed as a long-term replacement of petroleum for transportation.

The Japanese and Europeans, who were at one time behind the United States in fuel cell research, are now more advanced in their financial support to develop practical fuel cells.

EXAMPLES OF COMMON COMPOUNDS

Anhydrous ammonia (NH_3)—Colorless gas with a sharp, irritating odor. Lighter than air, easily liquified. An important commercial compound. First complex molecule found in space.

Hydrogen chloride and hydrochloric acid (HCl)—Toxic.

Hydrogen peroxide (H_2O_2)—An oxidizer, explosive.

Hydrogen fluoride (HF) and Hydrogen Bromide (HBr)—Both toxic.

Hydrogen sulfide (H_2S)—Rotten egg gas odor, poisonous.

Not so common: Heavy water (D_2O and T_3O)—Molecules composed of heavy hydrogen and oxygen are used as radioactive tracers in biochemical research, and as moderators in nuclear reactors.

Hydrocarbon compounds—Organic compounds such as: oil, gas, foods, animals, plants, etc. (some beneficial, some toxic).

HAZARDS

Hydrogen gas is very explosive when mixed with oxygen gas and touched off by a spark or flame. On May 6, 1937, the German rigid-frame dirigible or zeppelin, *Hindenburg*, which was inflated with hydrogen, exploded while landing at Lakehurst, New Jersey. The explosion killed thirty-six people and injured many more when the dirigible tried to land during a thunder and lightning storm.

Many hydrocarbon compounds are essential to life. On the other hand, many compounds containing hydrogen are poisonous and toxic.

❖ ❖ ❖

LITHIUM (Alkali Metals)
SYMBOL: Li   ATOMIC NUMBER: 3   PERIOD: 2

COMMON VALENCE: 1   **ATOMIC WEIGHT:** 6.939   **NATURAL STATE:** Solid **COMMON ISOTOPES:** There are several isotopes of lithium. The most common are lithium-6 and lithium-7.   **PROPERTIES:** Very soft silvery metal. Density: 0.534; melting point: 179°C; boiling point: 1340°C. About the same heat capacity as water. High electrical conductivity. Electropositive, but the least reactive with other elements of all the alkali metals.

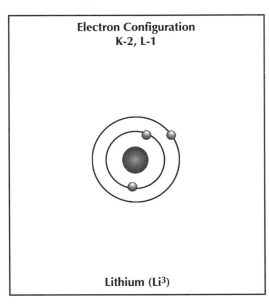

Electron Configuration
K-2, L-1

Lithium (Li³)

CHARACTERISTICS

Lithium is the lightest in weight of all metals.

Oddly, Li also exhibits properties of the alkali Earth metals in Group 2 (IIA).

It is an excellent reducing agent; i.e., it gives up electrons. Lithium will ignite and burn when exposed to oxygen and water. It is the only metal that reacts with nitrogen at room temperature.

Lithium has a high specific heat capacity and a great temperature range in liquid form, with very low density, which makes it a useful chemical.

ABUNDANCE AND SOURCE

Lithium is the thirty-first most abundant element. It does not exist in pure form and is always combined with other elements.

Lithium is contained in minute amounts in the minerals spodumene, lepidolite, and ambylgonite, which are found in the United States and several other countries in Europe, Africa, and South America. The Earth's crust contains just 65 parts per million (ppm) of lithium.

High temperatures are required to extract lithium from these minerals. It is also refined by solar evaporation of salt brine in lakes, and by **electrolysis** of lithium chloride.

HISTORY

Lithium was discovered in 1817 by Johan August Arfwedson (1792–1841). In 1818 the first lithium metal was prepared by Sir Humphry Davy (1778–1829). The name *lithium* is from the Greek word for "stone" because it is found in the Earth's minerals.

Lithium was discovered at about the same time in history when many "new" elements were being discovered and named. Many of these new elements were predicted by the use of the Periodic Table of the Chemical Elements.

COMMON USES

Lithium has many uses in today's industrial society.

It is used as a **flux** to promote the fusing of metals during welding and soldering. It also eliminates the formation of oxides during welding by absorbing impurities. This fusing quality is also important as a flux for producing ceramics, enamels, and glass.

A major use is as lithium stearate for lubricating greases. It makes a solid grease that can withstand hard use and high temperatures.

Lithium is used to manufacture storage batteries that have a long shelf life for use in heart pacemakers, cameras, etc.

Lithium is used for rocket propellants, nuclear reactor coolants, and alloy hardeners, and to make special ceramics.

Lithium is used as a source of alpha particles. When lithium nuclei (3 protons + 4 neutrons) are bombarded by accelerated protons (hydrogen nuclei), some of the Li nuclei break up into 4 protons + 4 neutrons, which, in turn, form two alpha particles (helium nuclei), e.g., $H^1 + Li^3 = He^2 + He^2$.

This is an example of the first man-made nuclear reaction produced by Cockcroft and Walton in 1929.

Pharmaceuticals: Several compounds of lithium are used to treat severe psychotic **depression** (as antidepressant agents). And, lithium carbonate is also used as a sedative or mild tranquilizer to treat less severe anxiety, which is a general feeling of uneasiness or distress about present conditions or future uncertainties. Lithium is used in the production of vitamin A.

EXAMPLES OF COMMON COMPOUNDS

There are numerous compounds of lithium. Its atoms combine with many other elements to form a variety of molecules. Some common examples are:

Lithium aluminum deuteride (LiAlD$_4$)—Used as source of deuterium atoms, which in turn produce tritium for nuclear reactors.

Lithium hydroxide (LiOH)—Storage batteries, soaps, carbon dioxide (CO$_2$) absorber in spacecraft.

Lithium aluminate (LiAlO$_2$)—Production of ceramics and porcelain enamels.

Lithium hydride (LiH)—Reducing agent, nuclear shielding, and as a **desiccant** to keep things dry.

Lithium fluoride (LiF)—Used to produce ceramics, as welding and soldering flux, rocket fuel, in light-sensitive scientific instruments, e.g., **X-ray diffraction** (scattering of X-rays by crystals that produce a specific pattern of that crystal's atoms, thus providing a technique for identifying different elements).

Lithium "salts" (such as LiCl)—Used for the treatment of different types of depression, especially manic depression.

HAZARDS

Lithium metal is highly flammable, explosive, and toxic. It will ignite in air, water, acids, or elements that are oxidizing agents and require electrons to complete their outer valence shell. It will even burn in nitrogen gas, which is relatively stable.

As an element (metal), it must be stored in oil or some air-free atmosphere. Many of its compounds will also burn when exposed to air or water. Lithium fires are difficult to extinguish, requiring special chemicals designed to extinguish them.

Solutions of lithium are very **toxic** to the human nervous system, thus requiring close observation by a physician when used as antidepressant drugs.

❖ ❖ ❖

SODIUM (Alkali Metals)
SYMBOL: Na   **ATOMIC NUMBER:** 11   **PERIOD:** 3

COMMON VALENCE: 1 **ATOMIC WEIGHT:** 22.989 **NATURAL STATE:** Solid **COMMON ISOTOPES:** No stable isotopes. Several rare isotopes exist, but are very radioactive with short half-lives. **PROPERTIES:** Sodium is a soft, wax-like silver metal that oxidizes in air. Density: 0.9674; melting point: 97.6°C; boiling point: 892°C. Sodium has excellent heat and electrical conducting qualities.

CHARACTERISTICS
On the Periodic Table sodium is located between lithium and potassium.

Sodium metal can be cut like a stick of butter. When cut it is silvery but turns gray as sodium oxide forms on the new surface.

Sodium is extremely reactive and will oxidize rapidly in air, but explosively in water as it releases hydrogen from the water. If not kept in an airtight container, it will ignite spontaneously with air. When in water, it forms hydrogen to produce sodium hydroxide.

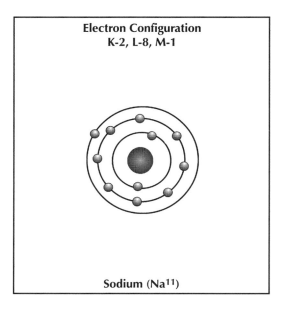

Electron Configuration
K-2, L-8, M-1

Sodium (Na¹¹)

ABUNDANCE AND SOURCE
Sodium is the sixth most abundant of the Earth's elements. It is not found in its pure elemental form on Earth, but it is relatively abundant. The Earth's crust consists of 2.83% of sodium found in the form of many different compounds. Commercially, sodium is one of the most important elements. After chlorine, sodium is the most abundant element found in solution in seawater.

HISTORY
Sir Humphry Davy (1778–1829) used an electric current to break up sodium carbonate to form metallic sodium. He first performed this experiment

with molten potassium carbonate to liberate the metal potassium, and soon followed up with the sodium experiment.

Today, sodium is still produced by the same **electrolytical** process, except that a mixture of sodium chloride and calcium chloride is used. It is then purified by **distillation**.

COMMON USES

There are numerous industrial, commercial, and household uses of sodium compounds. A few common examples are the following:

Sodium is used in both low-pressure and high-pressure sodium vapor lamps. The low-pressure arc uses just a small amount of Na and also some neon for a starter. The lamp is economical and bright, but its single yellow color makes it difficult to recognize other colors. The high-pressure lamp, in addition to sodium, uses mercury, which provides a more natural color rendition of light. The very bright light of sodium-mercury lamps makes them ideal for use in sports stadiums.

Soda niter or sodium nitrate is the most abundant of the nitrate minerals. It is used for fertilizers, explosives, and preservatives. The natural deposits are located in northern Chile, which was the original source for many years.

More recently, nitrogen fixation, which extracts nitrogen from the air, is used for producing sodium nitrate. This synthetic process eliminated the need for the natural source.

Of course, the most common use is everyday table salt, sodium chloride. Salt is vital for health; the body must have a small amount to survive. In the past, wars have been fought over salt mines and salt deposits by nations that did not have any natural sources. Excessive sodium chloride in the diet can also be harmful to one's health.

Many other useful sodium salts are sodium bicarbonate (baking soda), sodium carbonate (soda), sodium chloride (rock salt), sodium borate (borax), sodium sulfate (used in paper and photo industries). Mineral springs have a variety of sodium salts (and other trace compounds) that give the water its "fresh" taste.

EXAMPLES OF COMMON COMPOUNDS

We have mentioned only a few of hundreds of uses for sodium. Following are a few of the more common compounds of sodium:

Sodium chloride (NaCl)—Table salt.

Sodium thiosulfate ($Na_2S_2O_3 \bullet 5H2O$—Also known as "hypo," it is used to dissolve the unexposed silver salts from photographic negatives and prints during the process of "fixing" the image so the film or print will no longer be light-sensitive.

Sodium sulfite (Na_2SO_3)—As an **antioxidant**, it is used for a preservative except for meats. Used in water treatment, photographic developers, and textile bleaching.

Sodium sulfide (Na_2S)—Used in the dye industry, processing paper, sheep dip, engraving, the **oxidation** process of metal ores such as gold, lead, and copper.

Sodium silicate (Na_2O)—Also known as "water glass," it is used in making soaps and detergents, adhesives, bleach, water treatment, pigments, drilling fluids, etc.

Sodium permanganate ($NaMnO_4$)—Purple crystals, soluble in water. Used as an **oxidizing agent** disinfectant and bactericide. It can be used as an antidote for poisoning by morphine.

Rock salt—Unpurified, coarse common salt that is spread on highways to melt snow and ice.

HAZARDS

Sodium (elemental), as a metal, is very dangerous, particularly when in contact with air, water, snow, ice, or other active oxidizing agents. It releases hydrogen from the water with enough heat to ignite the hydrogen.

Sodium perchlorate ($NaClO_4$)—Dangerous, risk of fire and explosion. Used for explosives and jet fuel.

Sodium peroxide (Na_2O_2)—Risk of fire and explosion when in contact with water. Strong oxidizing agent, very irritating.

There are numerous sodium compounds that are hazardous as **carcinogens** (cancer causing) and as toxins to plants and animals. On the other hand, we benefit greatly from the many compounds containing the element sodium. We could not live without it.

POTASSIUM (Alkali Metals)
SYMBOL: K   ATOMIC NUMBER: 19   PERIOD: 3

COMMON VALENCE: 1 **ATOMIC WEIGHT:** 39.1 **NATURAL STATE:** Solid **COMMON ISOTOPES:** There are 3 natural isotopes of potassium: potassium-39, potassium-40, and potassium-41. Potassium-39 is not radioactive and is the most abundant. Only potassium-40 is naturally radioactive. There are many other man-made artificial radioactive isotopes of potassium. **PROPERTIES:** Potassium is a soft, silvery metal that oxidizes in air to form a dark gray potassium oxide coating. Density: 0.862; melting point: 63°C; boiling point: 770°C.

CHARACTERISTICS

Potassium is located in the fourth Period (row), in the middle of the alkali metals of Group 1 (IA).

It is a butter-like, lightweight metal with a low melting point. Potassium is more reactive than sodium, but similar to Na in its behavior. Potassium will rapidly oxidize in moist air if not kept in an oxygen-free atmosphere or stored in kerosene.

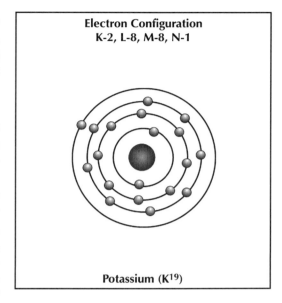

Electron Configuration
K-2, L-8, M-8, N-1

Potassium (K¹⁹)

ABUNDANCE AND SOURCE

Potassium is the seventh most abundant element in the Earth's crust (which contains 2.6% potassium). It is found as compounds (not "free") in all soils on Earth. Seawater contains 380 ppm of potassium in solution. Ore deposits are found in New Mexico, California, Salt Lake in Utah, Germany, Russia, and Israel.

Potassium is refined from sylvite, carnallite, and polyhalite ores. Potassium chloride is a common and useful raw material that is used to produce hundreds of other useful compounds. Potassium is produced by **thermochemical distillation** of potassium chloride.

HISTORY

In 1807, Sir Humphry Davy (1778–1829), using **electrolysis**, passed an electric current through melted potassium chloride (KCl). Small droplets of a silvery metal were produced, which reacted very strongly with water. Davy recognized that the potassium was reacting with the oxygen of water as hydrogen was released. The reaction was strong enough to raise the temperature to the degree that hydrogen was ignited.

COMMON USES

As with other alkali metals, potassium compounds have many uses. Some examples are:

Several compounds of potassium are used in fertilizers.

Explosives and fireworks use several types of potassium compounds.

Liquid potassium, when mixed with liquid sodium (NaK), is an alloy used as a heat exchange substance in nuclear reactors.

Potassium is an important **reagent** (something that is used in chemical reactions to analyze other substances) that forms many compounds used in chemical and industrial laboratories. Almost 1,000 chemical reagents are available for laboratory use in various types of reactions.

Potassium is used to manufacture both liquid and soft soaps.

It is an important raw material for making explosives, matches, and fireworks, e.g., gunpowder and the explosive in fireworks.

Potassium is essential to living organisms. It is a trace element required for a healthy diet and is found in many foods. One natural source is bananas.

EXAMPLES OF COMMON COMPOUNDS

There are hundreds of potassium compounds. Some common examples are:

Potassium nitrate (KNO_3)—Used in explosives (gunpowder), as a preservative, and in fertilizers. It is added to cigarette tobacco to make it burn evenly. It is used in glass and steel industries. KNO_3, also known as "saltpeter," has been used as a "sexual depressant," but the evidence for effectiveness is scanty.

Potassium chloride (KCl)—Used in drug preparations, photography, food additives, and chemical reagents. To reduce sodium in the diet, you can substitute potassium chloride for table salt. Also used for the production of metallic potassium.

Potassium carbonate (K_2CO_3)—Used for producing color TV tubes and other optical devices. Also useful as a food additive, for making paint and ink pigments, and as a **dehydrating** agent (a material that will "dry out" or remove water from substances).

Potassium chromate (K_2CrO_4)—Used to make bright yellow inks and paint pigments.

Potassium cyanide (KCN)—Used as an insecticide and fumigant. Used for electroplating and extracting gold and silver from ores. Also used as a source of cyanide (CN) gas in gas chambers.

Potassium bromide (KBr)—Used in photography and as a medical sedative.

Potassium iodide (KI)—Added to salt to help prevent enlargement of the thyroid gland, a condition known as a goiter.

HAZARDS

Potassium (elemental), as a metal, is very dangerous to handle. It can ignite while you are holding it with your hands as you cut it. It must be stored in an **inert** gas atmosphere or in oil. Potassium fires are extinguished by dry chemicals, such as soda ash or graphite. Water or regular fire extinguishers should not be used.

A particular hazard, which has been with humans since the beginning of time, is the radioactive isotope potassium-40. About 0.1% of all potassium on the Earth is in the form of this radioactive isotope. And there is no way for the body to distinguish the radioactive potassium from the nonradioactive form. Along with cosmic rays and other naturally radioactive elements, potassium-40 contributes to the normal lifetime accumulation of radiation. It makes up almost one-fourth of the total radiation our bodies receive over a normal life span.

Many of the potassium "salts" we mentioned are hazardous because they are explosive, either when heated or shocked. Some are also toxic to the skin and poisonous when **ingested.** On the other hand, numerous compounds of potassium make our lives much more livable. Potassium is essential to all life.

❖   ❖   ❖

RUBIDIUM (Alkali Metals)
SYMBOL: Rb   **ATOMIC NUMBER:** 37   **PERIOD:** 5

COMMON VALENCE: 1   **ATOMIC WEIGHT:** 85.47   **NATURAL STATE:** Solid
COMMON ISOTOPES: Rubidium-85.5 is the only stable form; rubidium-87 is the main natural radioactive isotope. **PROPERTIES:** Rubidium is a silvery-white, lightweight solid that is highly reactive. Density: 1.532; melting point: 39°C; boiling point: 688°C.

CHARACTERISTICS

Rubidium is located between potassium and cesium in the first Group in the Periodic Table.

It is the second most electropositive alkali element and reacts vigorously and explosively with air.

Rubidium produces very serious skin burns. It must be stored in kerosene.

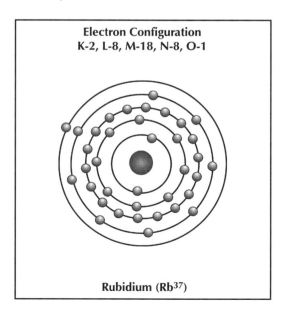

Electron Configuration
K-2, L-8, M-18, N-8, O-1

Rubidium (Rb^{37})

ABUNDANCE AND SOURCE

Rubidium is widespread and found in 310 ppm in the Earth's crust, making it the twenty-second most abundant element on Earth. It is found in many minerals along with other elements.

Seawater contains only 0.2 ppm, which is twice the amount of lithium found in the oceans.

Rubidium is similar to lithium and cesium, which are found in combined forms as complex minerals. It is not found in a pure elemental (metallic) state in nature, but only as compounds in mineral deposits such as pollucite, carnallite, and lepidolite, in addition to mineral springs.

Rubidium is produced by the thermochemical reduction of rubidium chloride (RbCl) with calcium (Ca). It can also be produced by **electrolysis** of melted cyanide (CN) or chloride (Cl).

HISTORY

The German scientist Gustav Robert Kirchhoff (1824–1887) discovered rubidium in 1860, along with cesium.

He experimented with the spectral lines of light, now called spectroscopic analysis or **spectroscopy**. While working with Robert W. Bunsen (1811–1899), who invented the gas Bunsen burner, they noted that, when heated or burned, different elements produced distinctive colors of light when transmitted through a **prism**. If the light was from the sun, dark lines for the elements in the sun were produced due to absorption by the sun's atmosphere. Since one element, when heated, produced a distinctive red line, Kirchhoff called it *rubidium* for the Latin word *rubidus*, which means "red."

COMMON USES

As an element, rubidium has limited use owing to its extremely electropositive nature (giving up its single, lightly held electron located in the outer shell, and thus becoming a positive ion).

Rubidium chloride is the source of most rubidium metal. Owing to rubidium's high heat transfer coefficient, the liquid form makes a good substance to transfer heat in nuclear reactors.

When rubidium gas is placed in a sealed glass cell along with an inert gas, it becomes a rubidium-gas cell clock. Because of the consistent and exact frequency (vibrations) of its atoms, it is a very accurate clock.

Rubidium is used as an **electrolyte** for low-temperature electric storage batteries for use in the polar regions of the Earth.

Rubidium, and also selenium, are used to make **photoelectric cells** (sometimes called *electric eyes*). When light strikes these elements, electrons are knocked loose from the outer shells of their atoms. These free electrons have the ability to carry an electrical current. Rubidium is also used as a **getter** to remove gases from vacuum tubes.

Photoelectric cells can be used to send a beam across a driveway or door, so that if the beam is broken an alarm sounds.

Rubidium makes an excellent **reducing agent** in both industry and chemical laboratories because of its strong electropositive nature.

A unique use is to locate brain tumors. It is a weak **radioisotope** that is able to attach itself to diseased tissue rather than healthy tissue, thus making detection possible.

EXAMPLES OF COMMON COMPOUNDS

A few useful compounds are:

Rubidium carbonate (Rb_2CO_3)—Used for making special glass.

Rubidium chloride (RbCl)—Source of rubidium metal and used as a chemical reagent.

Rubidium hydroxide (RbOH)—Is very **hygroscopic** (absorbs large amounts of water for its weight). It is also an excellent absorber of carbon dioxide. Rubidium is one of the few compounds that can be used to etch glass. It is also used for the electrolyte of low-temperature storage batteries.

HAZARDS

The major hazard is from fire and explosions of the metallic form of rubidium. It must be stored in an inert atmosphere or in kerosene.

If a piece of the metal gets on the skin, it keeps burning and produces a deep, serious wound. Water just makes it react more vigorously.

Most of the compounds of rubidium are toxic and strong irritants to the skin or when breathed. It is one of the elements best left to experienced users working in well-protected laboratory environments.

Very small traces of rubidium are found in the leaves of tobacco, tea, and coffee, as well as some other plants, but not enough to harm you if used in moderation.

CESIUM (Alkali Metals)
SYMBOL: Cs   **ATOMIC NUMBER:** 55   **PERIOD:** 6

COMMON VALENCE: 1 **ATOMIC WEIGHT:** 132.905 **NATURAL STATE:** Solid **COMMON ISOTOPES:** Cesium-133 is the only stable form of cesium occurring in nature. Cesium-137 is a radioactive isotope with a **half-life** of 33 years. It produces both **beta** and **gamma** rays. **PROPERTIES:** Cesium is the least electronegative and most reactive of the Earth metals (its single outer electron is very weakly attracted to its nucleus). Cesium metal is a soft solid at room temperature and turns to a liquid at slightly above room temperature. Density: 1.90; melting point: 28°C; boiling point: 705°C.

CHARACTERISTICS

Cesium is located between rubidium and francium in Group 1 of the Periodic Table.

Cesium is the heaviest of the stable alkali metals and is the one with the lowest melting point. It is also the most reactive of the alkali metals.

Cesium will decompose water, producing hydrogen, which will burn as it is liberated from H_2O. Cesium is extremely dangerous to handle and will burn spontaneously or explode when exposed to air, water, and many organic compounds. Since it may explode when exposed to oxygen, sulfur, phosphorus, and other oxidizing materials, it must be stored in an **inert** atmosphere.

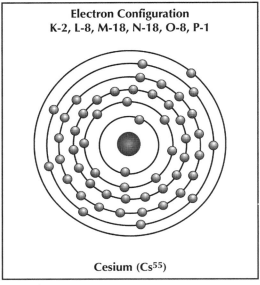

Electron Configuration
K-2, L-8, M-18, N-18, O-8, P-1

Cesium (Cs55)

ABUNDANCE AND SOURCE

Cesium is not very abundant; the Earth's crust contains only about 7 ppm. It is the forty-sixth most abundant element on the Earth.

Cesium is found in mixtures of complex minerals such as lepidolite ores. Lithium and rubidium are often found in the same minerals where cesium is found.

Its major ore is pollucite, which is found in Maine, South Dakota, Manitoba, Elba, and, mainly, Rhodesia in South Africa.

Cesium is also produced by the **thermochemical reduction** of a mixture of cesium chloride (CsCl) and calcium (Ca).

HISTORY

Cesium, originally spelled "caesium," was discovered in 1860 by Gustav Robert Kirchhoff (1824–1887), along with Robert W. Bunsen (1811–1899). They experimented with the spectrum of light passing through a **prism** (using **spectroscopic analysis, i.e., spectroscopy**) and noticed that each element, when burned, produced a distinctive color band of light after it passed through the prism.

The first new element identified by spectroscopic analysis was cesium; the second was rubidium. Kirchhoff coined the name *cesium*, which means "light-blue" or "bluish-green" in Latin.

COMMON USES

Cesium and its compounds have several important industrial uses. The following are some examples.

Because it emits electrons when exposed to light, it is used as a coating for photoelectric cells, scintillation counters (**Geiger counters**), and "electric eyes." It is used as a **getter** to absorb traces of oxygen and other gases in radio vacuum tubes.

Cesium is used as a **hydrogenation** catalyst (to enhance and assist the reaction) in the conversion of liquid oils to solid grease.

In a molten state, it is used as a heat transfer fluid in electric power generating plants.

Cesium is used as a **plasma** and ion source as a rocket propellant.

Cesium is used in military **infrared** devices and signal lamps, and other optical devices.

Cesium is used in the treatment of arsenic drug poisoning.

Cesium is used as a chemical **reagent** and **reducing agent** in industry and the laboratory.

Cesium-137 is a radioisotope that emits radiation at a very steady rate. This makes it useful as an atomic clock, as it is extremely accurate and never needs winding or a new battery!

EXAMPLES OF COMMON COMPOUNDS

Some common compounds are the following:

Cesium bromide (CsBr)—Used as crystals to produce **scintillation** (radiation) counters and fluorescent screens.

Cesium carbonate (Cs_2CO_3)—Used in brewing beer to make the "head" more foamy, in making specialty glass, and to make or enhance mineral water.

Cesium chloride (CsCl)—Used for mineral water, brewing, **fluorescent** screens, and production of other cesium compounds.

The radioactive isotope cesium-137 is used in industry as a source of strong but controllable radiation. It is used in medicine as a radiation source for cancer treatment to kill malignant cells. Cesium-137 is replacing cobalt-60 as a source of radiation in industry and medicine.

HAZARDS

As with other alkali metals, cesium is very dangerous to handle. Special precautions need to be taken to keep it away from air, water, snow, and organic substances with which it can vigorously react.

The radioisotope cesium-137 has a half-life of 33 years. During its decay it produces dangerous radiation that can cause serious radioactive poisoning. The Cesium-137 isotope is useful for sterilizing wheat, potatoes, and other foods. It is also useful in treating sewage sludge to kill harmful bacteria.

Cesium metal and many of its compounds are toxic, but many of its compounds are very useful to humans.

❖   ❖   ❖

FRANCIUM (Alkali Metals)
SYMBOL: Fr   ATOMIC NUMBER: 87   PERIOD: 7

COMMON VALENCE: 1 **ATOMIC WEIGHT:** 223 **NATURAL STATE:** Solid
COMMON ISOTOPES: No stable atoms of francium exist in nature. All isotopes are radioactive. The major isotope is actinium-K (francium-223), which is a decay product from the element actinium (with a half-life of about 21 or 22 minutes). Other isotopes are man-made. **PROPERTIES:** Francium's atoms are the largest and heaviest of the alkali metals (located just below cesium on the Periodic Table). Francium is a very weak electronegative element. Its chemical properties can only be determined by studying the radioactive decay process of actinium-K. Since it does have some properties similar to the alkali Earth metals, it fits best in the Periodic Table in Group 1 (IA). Not much is known about francium.

CHARACTERISTICS

No stable forms of francium have been found in nature.

Its nucleus is unstable with a short half-life of about 21–22 minutes.

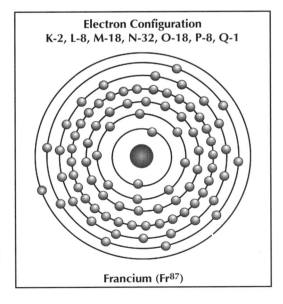

Electron Configuration
K-2, L-8, M-18, N-32, O-18, P-8, Q-1

Francium (Fr[87])

It is the only element from atomic numbers 1 to 92 that has a half-life of less than 30 minutes.

ABUNDANCE AND SOURCE

Francium is not a stable element; only traces have been found in nature. A very few francium salts exist. Those that do exist are water soluble. Because of the very short half-lives of its radioisotopes, only a few grams exist at any one time in the Earth's crust.

HISTORY

Marguerite Perey (1909–1975), a French physicist, discovered francium in 1939 while working with the radioactive element actinium in the Curie laboratory in the Radium Institute in Paris. As actinium-227 decayed, she tracked the end product to a new radioactive element, which she called francium-223 after her country—France.

COMMON USES

Since all forms of francium are radioactive, there are few practical uses for it—except as a source of radiation to study the radioactive decay process.

EXAMPLES OF COMMON COMPOUNDS

No useful ones exist.

HAZARDS

Decay of actinium-227 to francium-223 is a radiation hazard.

Group 2 Elements
THE ALKALI EARTH METALS

The alkali Earth metals are the elements in Group 2 (IIA) from Periods 2 to 7. They are: beryllium (Be^4), magnesium (Mg^{12}), calcium (Ca^{20}), strontium (Sr^{38}), barium (Ba^{56}), and radium (Ra^{88}).

They are not found as free elements in nature (only as compounds, in minerals and ores). They are more dense, less reactive, less volatile, and harder than the alkali metals. They do not burn in air when exposed at room temperature, but they do burn in air when heated. They have higher melting and boiling points than the metals in Group 1. Each of the alkali Earth metals gives off its own distinctive brilliant color when burned, which makes them useful for fireworks. Most of them are **malleable**, which means they can be worked into different shapes.

❖ ❖ ❖

BERYLLIUM (Alkali Earth Metals)
SYMBOL: Be   **ATOMIC NUMBER:** 4   **PERIOD:** 2

COMMON VALENCE: 2   **ATOMIC WEIGHT:** 9.012   **NATURAL STATE:** Solid
COMMON ISOTOPES: All isotopes are unstable. Historically, beryllium was also known as glucinum.   **PROPERTIES:** Beryllium is a grayish-white, hard, brittle metal. Density: 1.85; melting point: 1280°C. Soluble in all acids except nitric acid. Resists corrosion at room temperature. Has high heat capacity and transfer properties.

CHARACTERISTICS

Beryllium is the first alkali Earth metal in Group 2 (IIA) of the Periodic Table.

Beryllium has one-third the density of aluminum, and it resists oxidation, as does aluminum.

Beryllium metal can be machined (rolled, stretched, or pounded) and used in alloys to produce lightweight structural metals.

ABUNDANCE AND SOURCE

Beryllium is the forty-seventh most abundant element and is considered a rare metal. Its ore, called beryl, is found in South Africa, South America, and India, as well as in Colorado, Maine, New Hampshire, and South Dakota in the United States. Some deposits have been discovered in Canada.

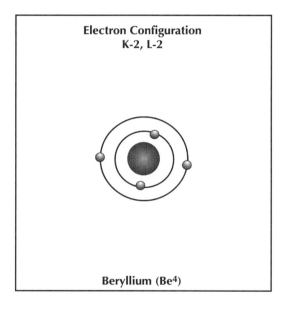

Electron Configuration
K-2, L-2

Beryllium (Be^4)

The beryl ore is oxidized, then converted to a **halogen**, such as chlorine or fluorine (Group 7 [VIIA]), and the metal is finally extracted by **electrolysis**.

HISTORY

In 1798 the French chemist Louis-Nicholas Vauquelin (1763–1829) discovered beryllium while he was in the process of identifying chromium. His sources were the gemstones beryl and emeralds.

Until the 1950s beryllium was known as glucinium, which was named for the Greek word *glukus*, meaning "sweet," because some of the solutions of beryllium tasted sweet.

The beryllium metal can be extracted from its ore by using high heat in a furnace, using magnesium metal as a **catalyst**, or by electrolysis.

COMMON USES

Because of its light weight, malleability, and other unique properties, beryllium is an excellent element for the manufacture of many items. Some examples are:

Beryllium is used in the manufacture of lightweight, but strong, space satellites, aircraft parts, and spacecraft.

As a 2% mixture with copper, it becomes an **alloy** that is about six times stronger than copper. This alloy is a better conductor of heat and electricity than most metals. Tools made from this alloy do not produce sparks when striking steel.

It is an excellent **moderator** to slow down high speed neutrons in nuclear reactors, and it acts as a reflector as well.

Beryllium is an excellent source of alpha particles, which are the nuclei of helium atoms. Alpha particles (radiation) are not very penetrating. These nuclei travel only a few inches in air and can be stopped by a sheet of paper. Alpha particles are produced in **cyclotrons** (atom smashers) and are used to bombard the nuclei of other elements to study their characteristics.

Beryllium was used as coating inside **fluorescent** electric light tubes, but proved to be **carcinogenic** (causes cancer) when broken tubes produced beryllium dust that was inhaled. Because of this potential to cause cancer, beryllium is no longer used as the inside coating of fluorescent tubes. Beryllium is also used for computer parts, electrical instrument components, and solid propellant rocket fuels. Since it is one of the few metals that is transparent to X-rays, it is used to make special glass for X-ray instruments and tubes.

EXAMPLES OF COMMON COMPOUNDS

There are many useful compounds of beryllium. Some examples are:

Beryllium carbide (Be_2C)—Used for the cores in nuclear reactors.

Beryllium chloride ($BeCl_2$)—Used as a catalyst to accelerate many organic chemical reactions.

Beryllium copper—Not really a compound, but a very useful alloy that often contains other metals such as cobalt or nickel in small amounts. It is a hard, strong alloy with excellent electrical conductivity, which makes it very useful in electrical switching equipment owing to its nonsparking qualities. It makes excellent spot-welding electrodes, springs, and metal bushings, cams and diaphragms.

Beryllium fluoride (BeF$_2$), also Beryllium sodium fluoride—Both used to produce the alkali Earth metal beryllium by reduction with magnesium metal.

Beryllium hydride (BeH$_2$)—When reacting with water or alcohol, hydrogen is liberated. Experimental use as rocket fuel.

HAZARDS

The elemental metallic form of beryllium, as well as most of its compounds, is highly **toxic**. When **inhaled**, the fumes, dust, or particles of beryllium are highly carcinogenic. Some beryllium compounds are toxic when they get into cuts in the skin, e.g., when an old fluorescent tube breaks.

Beryllium and its compounds should be handled *only* by experienced workers or laboratory personnel in proper facilities.

As with many other chemicals, there are good and bad sides to their use. Beryllium is a very important industrial element, and when its dangers are understood, it is safe to use.

❖ ❖ ❖

MAGNESIUM (Alkali Earth Metals)
SYMBOL: Mg   ATOMIC NUMBER: 12   PERIOD: 3

COMMON VALENCE: 2   **ATOMIC WEIGHT:** 24.31   **NATURAL STATE:** Solid

COMMON ISOTOPES: Magnesium has 3 rare isotopes. None are important.

PROPERTIES: Magnesium is a lightweight, silvery alkali earth metal. It is a strong electropositive reducing agent (gives up electrons). Density: 1.74; melting point: 650°C; boiling point: 1107°C.

CHARACTERISTICS

Magnesium has about 66% of the density of aluminum. It is placed as the second alkali Earth metal in Group 2 (IIA).

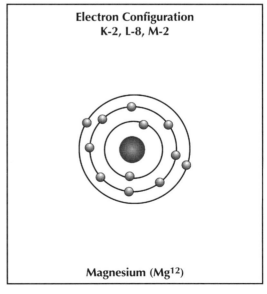

Electron Configuration
K-2, L-8, M-2

Magnesium (Mg^{12})

In solid metal and powder forms it ignites at 650°C to produce a brilliant white light.

It reacts with water, air, or alkalies at room temperature. Magnesium makes an excellent structural metal when alloyed with other metals. It is one of the lightest of all construction metals. It is **machinable** (can be rolled, pounded, and cut on a lathe).

ABUNDANCE AND SOURCE

Magnesium is the seventh most abundant element—it is not considered rare. Magnesium occurs in several mineral forms, including magnesite, serpentine, and dolomite. It is also found in seawater and salt beds.

After aluminum and iron, it is the most abundant structural metal on Earth.

Magnesium is produced by **electrolysis** of magnesium chloride ($MgCl_2$) from seawater, and by the reduction of magnesium oxide (MgO).

HISTORY

In 1755, Joseph Black (1728–1799) separated magnesium oxide from lime, but he did not separate the magnesium from the oxygen. Magnesium was identified as an element in 1808 by Sir Humphry Davy (1778–1829). Davy also discovered several other elements (e.g., potassium, barium, calcium, and strontium) by isolating these metals through electrolysis. Using this process he also discovered magnesium. He noticed that they all combined firmly with oxygen and could not be separated easily.

This led to the development of **electrochemistry**, which is the use of electricity as the energy source to break up the oxides of these elements.

COMMON USES

Magnesium and its alloys and compounds have many industrial, military, and medical uses. A few examples are:

Magnesium/aluminum alloys are used in producing lightweight structural parts for the space, the auto, and other industries.

In powder or foil form, magnesium burns with a bright white light and is used for fireworks and flash photography. In these forms it can be ignited by an electric spark or hot wire.

It can cause extensive fires as magnesium flares and firebombs used in wartime. Magnesium fires cannot be extinguished by water—dry sand or dirt is required.

It is used for the production (thermal reduction) of other metals, such as zinc, iron, titanium, zirconium, and nickel.

Owing to its strong **electropositive** nature (gives up electrons), it is used to "desulfurize" molten iron to produce steel by combining with the sulfur impurities in the iron to form MgS.

It is an important catalyst used for the synthesis of complex organic compounds.

Magnesium is an important element that acts as a catalyst (a substance that either speeds up or slows down a chemical reaction) in many "life" processes, e.g., **photosynthesis**, the oxidation process in our cells, and as a substance in red blood cells. We cannot live without it.

EXAMPLES OF COMMON COMPOUNDS

Many magnesium compounds are used in industry and agriculture. Some examples are:

Magnesium acetate [$Mg(OOCCH_3)_2$]—Used in the textile industry as a dye **mordant** ("fixes" dyes so they won't run). It is also used as a deodorant and antiseptic.

Magnesium chloride ($MgCl_2$)—Many uses, including source of magnesium metal, ceramics, lubricants, paper and textile manufacturing, disinfectants, fire extinguishers, and as a catalyst.

Magnesium fluoride (MgF_2)—Used to **polarize** corrective lenses of eyeglasses to reduce the glare of sunlight by changing the orientation of the light waves passing through the glass. Mg is also used to polarize windows, sunglasses, and similar items.

Magnesium chromate ($MgCrO_4$)—Medical uses, including dietary supplements and laxatives.

Magnesium oxide (MgO)—As a lining for steel furnaces, in ceramics, strong windows, fertilizers, paper and rubber manufacturing, food additives and **pharmaceuticals**. MgO_2 is also used as an antacid, e.g., milk of magnesia.

HAZARDS

Magnesium metal, particularly in the form of powder or small particles, can ignite easily and is very difficult to extinguish. Some form of dry dirt or sand, instead of water, must be used to extinguish magnesium fires. Magnesium will react with water to release hydrogen that will just intensify the fire.

Magnesium compounds, whose molecules include several atoms of oxygen, e.g., [$Mg(ClO_4)$], can be extremely explosive when in contact with organic compounds or moisture.

Many compounds of magnesium are poisonous when **ingested**.

❖   ❖   ❖

CALCIUM (Alkali Earth Metals)
SYMBOL: Ca   **ATOMIC NUMBER:** 20   **PERIOD:** 4

COMMON VALENCE: 2   **ATOMIC WEIGHT:** 40.08   **NATURAL STATE:** Solid
COMMON ISOTOPES: There are 6 stable isotopes of calcium, plus man-made radioactive calcium-45, which emits beta particles (high-speed electrons) and has a half-life of 164 days. This radioactive isotope is used to study the calcium content of bones, soils, etc.   **PROPERTIES:** Calcium is a soft, silvery, crystal alkali Earth metal. It will react with water to release hydrogen. Ca is an important element for the nutrition of living organisms. Vitamin D (and milk) aid in the deposition of calcium in bones and teeth. Density: 1.57; melting point: 845°C; boiling point: 1480°C.

CHARACTERISTICS

Calcium reacts with water to release hydrogen, and in powder form it is flammable in moist air.

Calcium is located between magnesium and strontium in Group 2 (IIA) of the Periodic Table.

It is harder than sodium, but softer than aluminum. In the metallic form it can be **machined** (cut on a lathe), **extruded** (pushed through a die), and drawn (stretched into rods and wire).

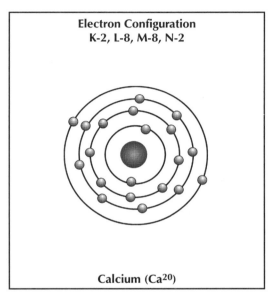

Electron Configuration
K-2, L-8, M-8, N-2

Calcium (Ca^{20})

Calcium is an essential **inorganic** element for plant and animal life.

ABUNDANCE AND SOURCE

Calcium is the fifth most abundant element on Earth. It is not found as a **free element**, but as calcium compounds (salts), which are found on all land masses of the world as limestone, marble, and chalk. Calcium, particularly as the compound calcium chloride ($CaCl_2$), is found in the oceans to the extent of 0.15%.

Calcium may be produced by the **electrolysis** of fused calcium chloride or by **thermochemistry** from lime and aluminum. As an elemental metal, calcium does not occur free in nature.

It also exists in bones, teeth, coral, and shells of some animals, including eggshells. It is an important element for healthy plants as well as animals.

HISTORY

In the year 1808, Sir Humphry Davy (1778–1829) experimented with electrolysis to separate several elements from their compounds that contained oxygen. By using this new process he discovered barium, strontium, and magnesium, as well as the element calcium.

COMMON USES

Calcium oxide was used in ancient times to make mortar for building with stones.

Both the metal and calcium compounds have many industrial as well as biological uses.

Metallic calcium is used as an **alloy** agent for copper and aluminum. It is also used to purify lead, and is a **reducing agent** for beryllium.

It is used to remove carbon and sulfur impurities from iron, producing a higher-grade iron for use in the manufacture of steel.

It is also used as a reducing agent in the preparation of several important metals.

Calcium is an important ingredient of the diets of all plants and animals. It is found in animal soft tissues and fluids, e.g., blood, as well as in bones and teeth. Calcium makes up about 2% of our body weight.

Ca is the main ingredient of Portland cement and is used to reduce the acid content of soils.

EXAMPLES OF COMMON COMPOUNDS

Many calcium compounds are found on both land and sea. A few important examples are:

Calcium acetate [$Ca(CH_3COO)_2$]—Many uses as food additives, as a **mordant** in the textile industry (to fix dyes so they won't run), as an alkali in the production of soaps and **lubricants**, in bakery goods (antimold), and for the manufacture of acetone.

Calcium bromide ($CaBr_2$)—Used in photography (as a developer), in medicines, and as a dehydrating agent (drying agent), food preservative, and fire retardant.

Calcium carbide (CaC_2)—Used to generate acetylene gas for welding and as a reducing agent. Explosive when in contact with water.

Calcium chloride ($CaCl_2$)—One of our most useful compounds. An excellent de-icer and road-dust reducer. Excellent drying agent, and used in the paper and pulp industries.

Calcium hydroxide [$Ca(OH)_2$]—Many uses. A major ingredient of mortar, plaster, cements, and soil conditioners. Used as a water softener, food additive, and **depilatory**.

HAZARDS

The metallic form of calcium, particularly the powdered form, combines with water or other **oxidizing** agents to release hydrogen, as do many of the other alkali metals. The elemental form, as well as some of the compounds, can be explosive, and there is the danger of fire when working with some calcium compounds.

The radioactive isotope calcium-45 is deposited in bones and teeth as well as other plant and animal tissues. Our bodies cannot tell the difference between stable and radioactive elements.

It is used as a **tracer** to study diseased bone and tissue. Since calcium-45 can displace the stable form of calcium-40, it can cause radiation sickness and even death.

A few calcium compounds, when in dust or powder form, are toxic when **ingested** or **inhaled**.

❖ ❖ ❖

STRONTIUM (Alkali Earth Metals)
SYMBOL: Sr   ATOMIC NUMBER: 38   PERIOD: 5

COMMON VALENCE: 2   **ATOMIC WEIGHT:** 87.62   **NATURAL STATE:** Solid
COMMON ISOTOPES: Four stable isotopes. Two important radioactive isotopes are strontium-89 and strontium-90. Other man-made radioisotopes of strontium have been artificially produced. **PROPERTIES:** Strontium is a pale-yellow, lightweight alkali Earth metal—very similar to calcium. It decomposes on contact with water to release hydrogen. Density: 2.54; melting point: 752°C; boiling point: 1390°C.

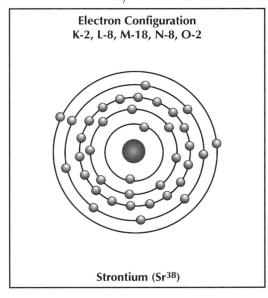

Electron Configuration
K-2, L-8, M-18, N-8, O-2

Strontium (Sr³⁸)

CHARACTERISTICS

It is located as the fourth element in Group 2 (IIA) of the Periodic Table.

Strontium can ignite when heated above its melting point and oxidizes quickly in air and water.

Metallic strontium must be stored in an **inert** atmosphere or in naphtha.

ABUNDANCE AND SOURCE

Strontium is never found in a free natural state on the Earth. It is the sixteenth most abundant element and the least abundant of the alkali Earth metals. Strontium exists as only 0.042% of the Earth's crust.

It is found in Mexico and Spain as the ores strontinite ($SrCO_3$) and celestite ($SrSO_4$).

It is produced by **electrolysis** of melted strontium chloride ($SrCl_2$) and by the reduction of strontium oxide by aluminum.

HISTORY

William Cruikshank (1745–1800) isolated, but did not identify, strontium from the mineral strontinite in 1787. Adair Crawford (1748–1794) gave strontium its name in 1790. Sir Humphry Davy (1778–1829) discovered strontium in 1808 at about the same time that he identified barium, magnesium, and calcium. Davy used **electrolysis** to separate these alkali Earth metals from their ores.

COMMON USES

Strontium does not have as many practical uses as do some of the other alkali Earth metals.

Strontium is used in making specialty metals when alloyed with other metals.

Strontium nitrate [$Sr(NO_3)_2$], when burned, produces a bright-red flame. It is used in fireworks, and during war it is used to make "tracer bullets" so their paths can be tracked at night.

It is used in making soaps, greases, and similar substances that are resistant to high and low temperatures.

EXAMPLES OF COMMON COMPOUNDS

Some useful compounds are:

Strontium-90. Radioactive strontium isotope with a half-life of 38 years. Dangerous, but useful in devices that measure the thickness or density of products by how transparent they are to radiation, e.g., cigarettes, building materials, and textiles.

Strontium hydroxide [$Sr(OH)_2$]—Used to extract sugar from sugar beet molasses. Also used to make soaps, adhesives, plastics, glass, and both high- and low-temperature lubricants.

Strontium iodide (SrI_2)—Used as the source of iodine.

Strontium nitrate [$Sr(NO_3)_2$]—Used in fireworks, matches, marine signals, etc.

Strontium sulfide (SrS)—Used as a depilatory (removes hair from skin surface). Smells like rotten eggs.

Many other strontium compounds are used in industries that produce paint, ceramics, glass, paper, and pharmaceutical products, including antiseptics.

HAZARDS

In powder form, strontium metal may spontaneously burst into flames. Both the metal and some of its compounds will explode when heated. Some of the compounds will explode if struck with a hammer.

The metal and a few strontium compounds will react with water to release hydrogen (as the strontium combines with the oxygen). The heat of the reaction can cause the hydrogen to burn or explode.

Some compounds, such as strontium chromate and strontium fluoride, are **carcinogens** and **toxic** if **ingested**.

The radioactive isotope strontium-90 is very dangerous, since it is a "bone seeker" and replaces calcium in bone tissue. Excessive exposure causes destruction of the tissue in bone marrow that produces the red blood cells in our bodies. Radiation poisoning and death may result.

Strontium-90, as well as other radioisotopes, is produced by nuclear explosions where these radioactive chemicals are transported in the atmosphere, and then ingested and inhaled by plants and animals. Once this became known, it led to the ban on the atmospheric testing of nuclear weapons.

BARIUM (Alkali Earth Metals)
SYMBOL: Ba   **ATOMIC NUMBER:** 56   PERIOD: 6

COMMON VALENCE: 2 **ATOMIC WEIGHT:** 137.34 **NATURAL STATE:** Solid **COMMON ISOTOPES:** Barium has 7 stable isotopes. **PROPERTIES:** Barium is a silver metal that is somewhat malleable (machinable, can be worked on a lathe, and by stretching, pounding, etc.). Density: 3.6; melting point: approximately 710°C; boiling point: approximately 1500°C. (These properties are difficult to accurately determine because barium is extremely reactive and will ignite in moist air.)

CHARACTERISTICS

Barium is the fifth element in Group 2 (IIA) of the alkali Earth metals.

Barium is very reactive with water, ammonia, oxygen, the halogens, and acids.

It is more reactive with water than are calcium and strontium, but less so than sodium.

In powder form it will burst into a bright-green flame at room temperature. It can be **machined** into various forms.

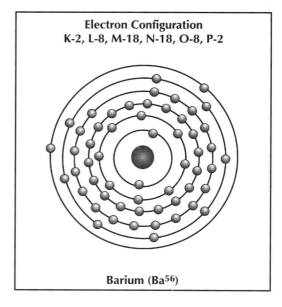

**Electron Configuration
K-2, L-8, M-18, N-18, O-8, P-2**

Barium (Ba⁵⁶)

ABUNDANCE AND SOURCE

Pure barium metal does not exist on Earth—only as compounds or in minerals and ores. Barium is the fourteenth most abundant element on Earth, making up about 0.05% of the crust.

It is found in the ores barite (barium sulfate) and barium oxide, also in witherite or barium carbonate. These ores are found in Missouri, Arkansas, Georgia, Kentucky, Nevada, California, Canada, and Mexico.

It is produced by the **reduction** of barium oxide (BaO) using aluminum or silicone in a high-temperature vacuum.

HISTORY

Carl Wilhelm Scheele (1742–1786) discovered barium oxide in 1774, but he did not isolate or identify the element barium.

It was not until 1808 that Sir Humphry Davy (1778–1829) used electricity to separate barium from its ore, by the procedure of **electrolysis**. Therefore, Davy received the credit for discovering barium.

COMMON USES

Barium and its compounds have many practical uses. Some examples are:

It is used in radio vacuum tubes, X-ray tubes, electronics, and spark plugs.

It is used in paint pigment, **lubricants**, refining edible oils, and ceramics.

It is used in experimental rocket fuels, fireworks, and as an analytical agent.

It is used in **insecticides**, **fungicides**, and rodent control. As a diagnostic, barium sulfate is swallowed by patients before being X-rayed. It is opaque to the X-rays and highlights the digestive system.

EXAMPLES OF COMMON COMPOUNDS

Many of the compounds of barium have similar practical uses as do strontium, calcium, and magnesium. Several examples of useful compounds of barium are:

Barium acetate [$Ba(C_2H_3)_2)_2$]—Produced by adding acetic acid to barium sulfate and recovered as crystals by evaporation. Used as dryers for paints and varnishes.

Barium bromate [$Ba(BrO_3)_2$]—An excellent **corrosion** inhibitor. Used as a chemical **reagent** and **oxidizing** agent.

Barium chloride ($BaCl_2$)—Used to make other barium chemicals, and in manufacture of pigments, additives for oils, and dyeing textiles.

Barium hydrate [$Ba(OH)_2$] (all forms)—Many uses, e.g., oil and grease additives, water treatment, and **vulcanization**. Used in the manufacture of soaps, beet sugar, glass, and steel.

Barium sulfate ($BaSO_4$)—Medical use as an opaque liquid medium to block out X-rays (shows up dark in diagnostic tests of the digestive system). Barium sulfate medium assists, when using X-rays, in providing an image of ulcers and intestinal problems. Also used in manufacturing inks, rubber, plastics, and paints.

Barium thiosulfate (BaS_2O_3)—Used in explosives, matches, photography, and varnishes.

HAZARDS

Barium metal, in powder form, is **flammable** at room temperature. It must be stored in an oxygen-free atmosphere or in petroleum.

Many of the barium compounds are also explosive (used in **pyrotechnics**, e.g., green-colored fireworks) and are **toxic** if **ingested** or **inhaled**. Care should be used when working with barium and other alkali metals in the laboratory or in industry.

RADIUM (Alkali Earth Metals)
SYMBOL: Ra   **ATOMIC NUMBER:** 88   **PERIOD:** 7

COMMON VALENCE: 2 **ATOMIC WEIGHT:** 226.03 **NATURAL STATE:** Solid **COMMON ISOTOPES:** There are many isotopes of radium, ranging from radium-206 to radium-234. All are radioactive. Radium-226 is the most abundant and most important of the isotopes. It has a half-life of 1620 years. **PROPERTIES:** Radium is a bright white radioactive luminescent alkali Earth metal that turns black when exposed to air. Radium decays by emitting three

types of radiation, i.e.,
alpha, **beta** and **gamma**
radiation. Density: 5.0;
melting point: 700°C;
boiling point: 1140°C.

Electron Configuration
K-2, L-8, M-18, N-18, O-32, P-8, Q-2

Radium (Ra^{88})

CHARACTERISTICS

Radium is the last element in Group 2 (IIA). It is similar to the other alkali Earth metals in its chemistry, particularly the element barium.

It is chemically reactive and forms many compounds.

Radium is an extremely dangerous element. Its radioisotopes seek out bone tissue if **ingested**. It is highly **toxic**.

ABUNDANCE AND SOURCE

Radium is the eighty-fifth most abundant element on the Earth. It is widely distributed, but not highly concentrated, in the Earth's crust and rocks. Radium is found in uranium minerals to the extent of only one part to every 3 million parts of uranium.

Radium is found in the ores **pitchblende** and carnotite. It is also isolated in small quantities from uranium metal.

Ore deposits are found in Colorado, Canada, France, Russia, and West Central Africa (Zaire). Metallic radium can also be separated from radium salts by **electrolysis**.

HISTORY

In the late 1890s Marie Sklodowska Curie (1867–1934) isolated radium in very small amounts from uranium ore (pitchblende). The chemical processes she used required much persistence.

Her husband, Pierre Curie (1859–1906), discovered a process called **piezoelectricity**, which was used by Madame Curie to measure the intensity of the radiation of the new element she discovered. Their assistant, G. Bemont, aided the Curies with the task of reducing tons of pitchblende ore to recover only a few grams of radium.

COMMON USES

Radium's most important use is as a source of radiation in industry, medicine and the laboratory.

The isotope radium-226, which has a half-life of 1620 years, is the only really useful form of all the isotopes. It is used in the medical treatment of malignant cancer growths. It kills cancer cells that have spread throughout the body.

Other uses are to produce **phosphorescence** and **fluorescence** in organic compounds and for **scintillation** screens on instruments used to detect radiation. Radium salts were used in the past to paint the dials of **luminous** clock faces that glow in the dark.

Radium is also the source of neutrons to bombard other elements and is the source of the element radon, which is one of radium's radioactive decay products.

EXAMPLES OF COMMON COMPOUNDS

Radium bromide ($RaBr_2$)—Used in cancer treatments and particle research in fields of physics.

Radium chloride ($RaCl_2$)—Same uses as for radium bromide.

HAZARDS

Radium is very toxic. The ionizing radiation can destroy plant and animal cells. If ingested, radium can replace bone tissue and cause severe radiation sickness or death.

The dry salts of radium are kept in sealed glass tubes, which, in turn, are stored in lead containers. Radium must be handled by experienced personnel and with caution.

Many years ago women painted clock and watch dials with luminous radium paint (a mixture of radium salts and zinc sulfide). They would "point" their paint brushes with their lips and tongue and then dip their brushes into the paint. In time, most of them developed badly eaten away and disfigured jaws, mouth tissue, and teeth. Many died prematurely of radiation poisoning and cancer. Once the dangers were known, the practice of painting luminous clock and watch dials was accomplished by using promethium, a less dangerous radioactive element.

5

TRANSITION ELEMENTS: METALS TO NONMETALS

The **transition elements** form three series of metals that progress from elements that give up or lose electrons (metals), to elements that gain or accept electrons (nonmetals). The elements that are in transition from metals to nonmetals are located in the center of the Periodic Table in Periods 4, 5, and 6, in Groups 3, 4, 5, 6, 7, 8, 9, 10, 11, and 12.

The transition metals are unique in that they represent a gradual shift on the **electronegativity** scale from very weak electronegative metallic elements to stronger electronegative nonmetallic elements. The positive nucleus of neutral atoms that have less than eight electrons in their valence shells attracts more electrons. This attraction is called "electronegativity," which is a tendency for some elements to gain electrons. Nuclei of metal atoms have a weak attraction for more electrons and exhibit the least electronegativity. In other words, transition elements range from elements with weak electronegativity to nonmetallic elements whose atoms are more receptive to gaining electrons and, thus, exhibit greater electronegativity.

← Weak Electronegativity ⟷ Strong Electronegativity→

Groups 1 & 2   ← Transition Metals →   Groups 13, 14, 15, 16, 17

The transition metals are also unique in that their outermost **valence** shell is not the main energy level being completed. Generally speaking, the next to the outermost shell provides the electrons that make the atoms of the

PERIODIC TABLE OF THE ELEMENTS

GROUPS / PERIODS

1 IA	2 IIA	3 IIIB	4 IVB	5 VB	6 VIB	7 VIIB	8 — VIII	9 VIII	10 VIII	11 IB	12 IIB	13 IIIA	14 IVA	15 VA	16 VIA	17 VIIA	18 VIIIA
1 H 1.0079																	**2** He 4.00260
3 Li 6.941	**4** Be 9.01218											**5** B 10.81	**6** C 12.011	**7** N 14.0067	**8** O 15.9994	**9** F 18.9984	**10** Ne 20.179
11 Na 22.9898	**12** Mg 24.305											**13** Al 26.9815	**14** Si 28.0855	**15** P 30.9738	**16** S 32.066(6)	**17** Cl 35.453	**18** Ar 39.948
19 K 39.0983	**20** Ca 40.08	**21** Sc 44.9559	**22** Ti 47.88	**23** V 50.9415	**24** Cr 51.996	**25** Mn 54.9380	**26** Fe 55.847	**27** Co 58.9332	**28** Ni 58.69	**29** Cu 63.546	**30** Zn 65.39	**31** Ga 69.72	**32** Ge 72.59	**33** As 74.9216	**34** Se 78.96	**35** Br 79.904	**36** Kr 83.80
37 Rb 85.4678	**38** Sr 87.62	**39** Y 88.9059	**40** Zr 91.224	**41** Nb 92.9064	**42** Mo 95.94	**43** Tc (98)	**44** Ru 101.07	**45** Rh 102.906	**46** Pd 106.42	**47** Ag 107.868	**48** Cd 112.41	**49** In 114.82	**50** Sn 118.71	**51** Sb 121.75	**52** Te 127.60	**53** I 126.905	**54** Xe 131.29
55 Cs 132.905	**56** Ba 137.33	**57** La 138.906 ★	**72** Hf 178.49	**73** Ta 180.948	**74** W 183.85	**75** Re 186.207	**76** Os 190.2	**77** Ir 192.22	**78** Pt 195.08	**79** Au 196.967	**80** Hg 200.59	**81** Tl 204.383	**82** Pb 207.2	**83** Bi 208.980	**84** Po (209)	**85** At (210)	**86** Rn (222)
87 Fr (223)	**88** Ra 226.025	**89** Ac 227.028 ▲	**104** Und (261)	**105** Unp (262)	**106** Unh (263)	**107** Uns (264)	**108** Uno (265)	**109** Une (266)	**110** Uun (267)	**111** Uuu (272)	**112** Uub	**113** Uut	**114** Uuq	**115** Uup	**116** Uuh	**117** Uus	**118** Uuo

TRANSITION ELEMENTS

Lanthanide and Actinide series:

6 ★ Lanthanide Series (RARE EARTH)	**58** Ce 140.12	**59** Pr 140.908	**60** Nd 144.24	**61** Pm (145)	**62** Sm 150.36	**63** Eu 151.96	**64** Gd 157.25	**65** Tb 158.925	**66** Dy 162.50	**67** Ho 164.930	**68** Er 167.26	**69** Tm 168.934	**70** Yb 173.04	**71** Lu 174.967
7 ▲ Actinide Series (RARE EARTH)	**90** Th 232.038	**91** Pa 231.036	**92** U 238.029	**93** Np 237.048	**94** Pu (244)	**95** Am (243)	**96** Cm (247)	**97** Bk (247)	**98** Cf (251)	**99** Es (252)	**100** Fm (257)	**101** Md (258)	**102** No (259)	**103** Lr (260)

transition elements electropositive. These elements are the only ones that use the next to the outermost shell in bonding with nonmetals. All other metals and nonmetals in the major Groups use their outermost shells as their valence shells in bonding.

These characteristics make the transition metals extremely useful. They form a variety of different and complex ions which account for many of the unique properties of metals.

There are several ways to present the transition elements. We present them as three series found in Periods 4, 5, and 6.

The first series starts at Period 4, Group 3, with the element scandium (Sc^{21}). The second series starts at Period 5, Group 3, with the element yttrium (Y^{39}). The third series starts at Period 6. It follows a number of special **"rare-earth"** type of metals called the **lanthanide series**. We start the third series with the element hafnium (Hf^{72}).

Some authorities begin and end the transition elements at different Groups in the Periodic Table. The transition elements sometimes continue beyond Group 12 to include some metallic and semiconducting (metalloid) elements in Groups 14, 15, and 16. We will start each of the transition elements with Group 3 and continue through Group 12.

First Series
(PERIOD 4; GROUPS 3 TO 12)

SCANDIUM (Period 4)
SYMBOL: Sc   **ATOMIC NUMBER:** 21   **GROUP:** 3

COMMON VALENCE: 2 and 3 **ATOMIC WEIGHT:** 44.956 **NATURAL STATE:** Solid **COMMON ISOTOPES:** Isotopes range from scandium-40 to scandium-51. All are **radioisotopes** produced by nuclear reactions, except scandium-45, which occurs as the stable natural form. **PROPERTIES:** Scandium is a silvery-white metal that does not **tarnish** when exposed to air. It is electropositive (easily gives up its valence electrons). Transition elements are unique in that they can use electrons in other than their outer shell for bonding with other elements.

CHARACTERISTICS
Scandium is the first of the transition elements in Period 4, Group 3 (IIIB), of the Periodic Table.

Sc has some properties similar to the rare-earth elements, but is not considered a rare-earth element. Scandium reacts vigorously with acids.

ABUNDANCE AND SOURCE

Scandium is the thirty-fifth most abundant element found in the Earth's crust. It is widely distributed over much of the world. Scandium is found in ores of wolframite in Norway and thortveitite in Madagascar. It is also found in granite pegmatitites and monazites. It is common in many of the ores where tin and tungsten are also found. Much more scandium is found in the stars and our sun than on the Earth.

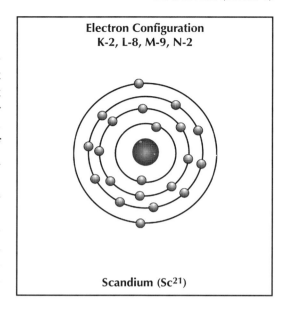

Electron Configuration
K-2, L-8, M-9, N-2

Scandium (Sc^{21})

HISTORY

In 1879, Lars Fredrik Nilson (1840–1899), of Sweden, discovered scandium (named for Scandinavia) while identifying several of the rare-earth elements. Scandium has some properties of the rare-earth elements, but is similar to other metals in several ways. Sc was the second new element predicted by Mendeleyev to fit into his Periodic Table. He called it "eka-boron" since he thought it would resemble boron once it was identified. Even though Mendeleyev had not discovered scandium, he knew what its physical and chemical properties should be. Some years later, Per Teodor Cleve (1840–1905) identified the properties of the new element as scandium. Cleve's discovery strengthened the acceptance of Mendeleyev's Periodic Table.

COMMON USES

Although there is no major industrial use for scandium, its oxides and some other compounds are used as **catalysts** to speed up chemical reactions. It is used in nickel alkaline storage batteries and as a radioactive **tracer**. It has some use in the space industry.

EXAMPLES OF COMMON COMPOUNDS

Scandium sulfate solution—Used to assist in the germination of seeds for agricultural plants.

Radioactive scandium-47—Used as a tracer in research. It is also used as a tracer in the petroleum industry.

HAZARDS

There are no hazards, except radiation from the radioactive isotopes.

TITANIUM (Period 4)
SYMBOL: Ti   **ATOMIC NUMBER:** 22   **GROUP:** 4

COMMON VALENCE: 2, 3, and 4 **ATOMIC WEIGHT:** 47.90 **NATURAL STATE:** Solid **COMMON ISOTOPES:** Five common isotopes are known. **PROPERTIES:** Titanium is a silver-gray powder that is as hard as steel but about 1/2 lighter. Titanium makes excellent **alloys** with iron and aluminum, which are light in weight and very resistant to seawater and chlorine **corrosion**. It is reactive and forms many useful compounds. Density: 4.6; melting point: 1675°C; boiling point: 3260°C.

CHARACTERISTICS

Titanium is the second element of Period 4 of the transition elements. It is similar to zirconium in both chemical and physical properties.

Titanium has many remarkable properties that make it useful in various industries.

ABUNDANCE AND SOURCE

Titanium is the ninth most abundant element. It is not found in pure form in nature but rather as **oxides** in titanite ore and the minerals rutile (TiO_2) and il-

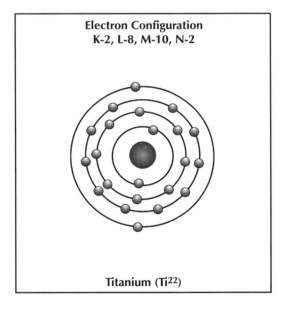

Electron Configuration
K-2, L-8, M-10, N-2

Titanium (Ti^{22})

menite ($FeTiO_3$). It is also found in some iron ores. Titanium is widely distributed over the Earth, but high concentrations are difficult to find. Titanium is also difficult to produce in highly refined, commercially pure forms (powder, metal forms, and crystals). It is found in the stars, our sun, the moon, and in meteorites.

HISTORY

In 1791, William Gregor (1761–1817) isolated an odd substance from several other minerals he was studying. He called it "menachite." During this period, Franz Joseph Muller (1740–1825) also produced a similar substance that he could not identify. Four years later Martin Heinrich Klaproth (1743–1817), who is considered the "father" of modern analytical chemistry, confirmed that this mineral was a new element. Klaproth named it *titanium*, which means "Earth" in Latin. The name *titanium* also refers to the titans of Greek mythology. Titanium was a difficult metal to extract from its ores.

COMMON USES

Because of titanium's lightness, strength, and resistance to corrosion and high temperatures, its most common use is in alloys with other metals for constructing aircraft, jet engines, and missiles. Its alloys also make excellent armor plate for tanks and warships.

Titanium's noncorrosive and lightweight properties make it useful for manufacturing laboratory and medical equipment that will withstand acid and halogen salt corrosion. Because it is light, strong, and resists corrosion, it is used to make pins and screws to repair broken bones.

It is also used for **abrasives**, to make cements and space-age batteries.

Titanium is used as a filler in paper and in **pigments** for ink and paint. It has many uses in a variety of industries.

EXAMPLES OF COMMON COMPOUNDS

Titanium reacts with many elements, including the halogens. It also absorbs hydrogen. A few examples of some compounds are:

Titanium dioxide (TiO_2)—Produced when titanium burns in air, titanium dioxide is known as rutile in its natural form. Titanium dioxide is used as a pigment for paints, especially paints exposed to severe temperatures and marine climates because of its inertness and self-cleaning attribute. It is used in the manufacture of glassware, ceramics, enamels, welding rods, and floor coverings.

Titanium bromide (TiB_2)—Used for **metallurgy**, high-temperature electrical wiring, electronics, computers, high-temperature-resistant coatings, and super alloys, including strong, lightweight aluminum alloys.

Titanium carbide (TiC)—Used as an additive to make high-temperature cutting tools, cements, and coatings.

HAZARDS

Titanium metal powder is highly **flammable** and is an explosion risk when heated in air. It will also burn in nitrogen. Titanium fires cannot be extinguished by using water or carbon dioxide. Sand, dirt, or special foams must be used to extinguish burning titanium.

VANADIUM (Period 4)
SYMBOL: V   ATOMIC NUMBER: 23   GROUP: 5

COMMON VALENCE: 2, 3, 4, and 5 **ATOMIC WEIGHT:** 50.9414 **NATURAL STATE:** Solid **COMMON ISOTOPES:** Two natural isotopes exist. Vanadium is not found in the natural state—only in mineral ores and in combination with other metals. **PROPERTIES:** Vanadium is a silvery/white, soft, **ductile** metal that can be worked into various shapes. Vanadium forms many complicated compounds due to its variable valences. In this respect, vanadium is similar to other transition metals where some electrons from the element's next-to-outermost shell can bond with other elements. Density: 6.11; melting point: 1900°C; boiling point: 3400°C.

CHARACTERISTICS

Vanadium is resistant to corrosion, and thus is suitable to make alloys for use in seawater. It is even more resistant to corrosion than stainless steel. When worked as a metal, it must be heated in an inert atmosphere because it will readily **oxidize**.

ABUNDANCE AND SOURCE

Vanadium is the nineteenth most abundant element. It is never found as a free metal in nature, but is

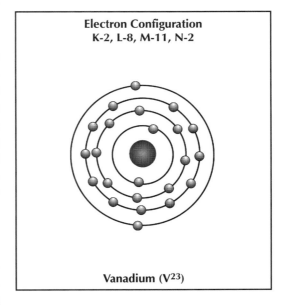

Electron Configuration K-2, L-8, M-11, N-2

Vanadium (V^{23})

quite common in a number of mineral ores. The main ores are carnotite, vanadinite, roscoelite, and patronite, found in several southwestern states of the United States, as well as in Mexico and Peru.

Vanadium is also recovered from the refining of phosphate rock during the production of phosphorus, mainly in the states of Arkansas, Montana, and Idaho.

HISTORY

Two unrelated discoveries of vanadium seem to have occurred. In 1830, Nils Gabriel Sefstrom (1787–1845), when experimenting with iron, identified a small amount of a new metal. Since vanadium compounds have beautiful colors, he named this new metal after *Vanadis*, the name for "Freya," the goddess of youth and beauty of his native country, Scandinavia.

Earlier, in 1801 in Mexico, Andres Manuel Del Rio discovered an unusual substance he called *erythronium*, but was told it was similar to chromium. It was only later that it was found to be vanadium. Some references give the credit for discovery to Del Rio, but the most recent references list Sefstrom as the discoverer.

Sometime later, in 1869, vanadium was isolated from its ores by Henry Enfield Roscoe (1833–1915), who also made many studies of the effects of light on substances. But, Sefstrom had already received credit for discovery of the element.

COMMON USES

A major use of vanadium is for special steel alloys. It makes tough, heat-resistant, strong steels that resist corrosion. Some of its compounds, particularly the oxides, are used in the chemical industry as **catalysts** to speed up organic chemical reactions.

It is also used as a photographic developer, to dye textiles, and in the production of sulfuric acid and artificial rubber.

When combined with glass, it acts as a filter to **ultraviolet** rays from sunlight.

EXAMPLES OF COMMON COMPOUNDS

A few common and useful compounds of vanadium are:

Vanadium pentoxide (V_2O_5)—A reddish-yellow powder extracted from minerals using acids or strong alkalies; used as a catalyst for many organic chemical reactions, to color ceramics, to dye cloth, in photography, and in UV-protected windowpanes.

Vanadium carbide (VC)—Used to produce alloys that are formed into high-speed, high-temperature tools for cutting metals and other substances.

Vanadium sulfate ($VOSO_4$)—Used as a catalyst for several important products, as a **reducing agent**, to color glass and ceramics, and as a **mordant** (binding a dye to textiles).

HAZARDS

Vanadium powder and dust and most of the compounds (oxides) of vanadium are toxic when inhaled.

Vanadium chloride compounds are also strong irritants to the skin and poisonous when ingested.

Many of its compounds must be stored in a dry, oxygen-free atmosphere or in containers of inert gas. Protective clothing and goggles should be worn when handling vanadium, as well as most of the other transition elements.

❖ ❖ ❖

CHROMIUM (Period 4)
SYMBOL: Cr   ATOMIC NUMBER: 24   GROUP: 6

COMMON VALENCE: 2, 3, and 6 **ATOMIC WEIGHT:** 51.996 **NATURAL STATE:** Solid **COMMON ISOTOPES:** Chromium-50, chromium-52, chromium-53, and chromium-54 are 4 natural, stable isotopes. There are several radioactive isotopes produced in nuclear reactors. **PROPERTIES:** Chromium is a silver-gray, brittle metal not found as a free element in nature. As a transition element, chromium can use electrons in its "M" shell as valence electrons when combining with other elements. Density: 7.1; melting point: 1900°C; boiling point: 2200°C.

CHARACTERISTICS

As a metal, chromium can be molded, **forged**, rolled, and drawn, but with difficulty unless it is in a very pure form.

Chromium is an excellent metal for making **alloys**. It is also used for electroplating due to its noncorrosiveness and its excellent reflective surface.

ABUNDANCE AND SOURCE

Chromium is the twenty-first most common element on Earth. It is not found as a

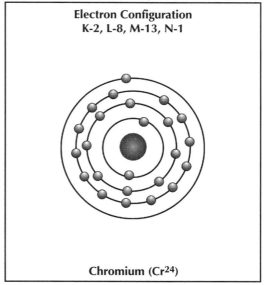

Electron Configuration
K-2, L-8, M-13, N-1

Chromium (Cr24)

free metal, but occurs only as compounds in nature. It is found in chromite mineral ores ($FeCr_2O_4$) located in Cuba, Zimbabwe, South Africa, Turkey, Russia, and the Philippines. There are different grades and forms of chromium ores, based on the classification of use of the element. Most oxides of chromium are found along with other metals, such as iron, magnesium, or aluminum.

Of some interest is that the astronauts found that the moon's basalt rocks contain several times more chromium than is found in basalt rocks on Earth.

HISTORY

In 1797 the French chemist Louis-Nicolas Vauquelin (1763–1829) discovered chromium while studying some minerals that were collected in Siberia. Its compounds produce many different colors, e.g., green, red, yellow, and silver, as well as different hues of these colors. Chromium was named for the Greek word *chromos*, the Greek word for "color."

COMMON USES

The best-known use of chromium is for the plating of metal and plastic parts to produce a shiny, reflective finish on automobile trim, household appliances and other items where a bright finish is considered attractive. It also protects iron and steel from corrosion.

It is used to make alloys, especially stainless steel for cookware and for items where strength and protection from rusting and high heat are important.

Its compounds are used for high-temperature electrical equipment, as a **mordant** (fixes the dyes in textiles so they will not run), for tanning leather, and as an antichalking agent for paints.

Some research has shown that, even though most chromium compounds are **toxic**, a small trace of chromium is important for a healthy diet for humans. A deficiency produces diabetes-like symptoms, which can be treated with a diet of whole-grain cereal, liver, and brewer's yeast.

Chromium's most important radioisotope is chromium-51, which has a half-life of about 27 days. It is used as a radioisotope **tracer** to check the rate of blood flowing in constricted arteries.

EXAMPLES OF COMMON COMPOUNDS

There are many useful compounds of chromium. A few examples are:

Chromium borides (CrB, CrB_2, and Cr_3B_2)—Used as metal additives, refractory coatings, and high-temperature conductors.

Chromic oxide (Cr_2O_3)—Used as a green paint pigment, in ceramics, in metallurgy, to make green asphalt roof shingles, and as a chemical catalyst. Chromic sulfate is used for similar purposes.

Chromic acid (CrO_3)—Exists only in solution as H_2CrO_4. Used for metal and plastic plating, engraving, anodizing, to color glass and glazes, and to make green paints and textiles.

HAZARDS

Most of the compounds of chromium are hazardous when inhaled and irritating when in contact with the skin. Even though chromium may be a necessary trace element in our diets, many of its compounds are very toxic when ingested.

Chromic acid is very hazardous. It will explode on contact with organic chemicals and is poisonous to humans and animals.

Chromite iron ore, which is found in **igneous rocks**, is **carcinogenic** and will cause lung cancer, even when a small amount of the ore dust is inhaled.

Workers in chromium-producing and -using industries are subject to bronchogenic cancer if precautions are not taken.

❖ ❖ ❖

MANGANESE (Period 4)
SYMBOL: Mn   **ATOMIC NUMBER:** 25   **GROUP:** 7

COMMON VALENCE: 2, 3, 4, 6, and 7 **ATOMIC WEIGHT:** 64.938
NATURAL STATE: Solid **COMMON ISOTOPES:** No stable isotopes are known. Four radioactive isotopes have been produced in nuclear reactors.
PROPERTIES: Manganese is a silvery, brittle, hard metal. There are four allotropic forms of manganese (allotropy means a different form of the same element with different chemical or physical properties). Density: 7.44; melting point: 1245°C; boiling point: 2097°C.

CHARACTERISTICS

Manganese is located between chromium and iron in the first row (Period 4) of the transition elements. It has characteristics of both its neighbors.

As a pure metal, it cannot be worked into different shapes, as it is too brittle.

Manganese is a reactive metal that has several oxidation states due to its mul-

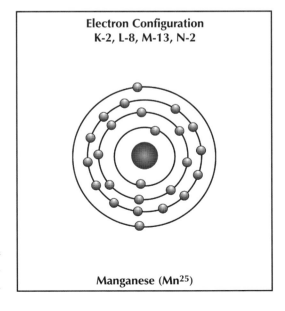

Electron Configuration
K-2, L-8, M-13, N-2

Manganese (Mn²⁵)

tiple valences. As a trace element it is essential for the well-being of plants and animals.

ABUNDANCE AND SOURCE

Manganese is the twelfth most abundant element on Earth. It is not found as a pure metal in nature. Pure manganese is found in meteorites that land on the Earth's surface. It is often located in low-grade iron ores and in mineral ores such as psilomelane, pyrolusite, rhodichrosite, and manganite (manganese ore). These ores are found in India, Brazil, Montana, and areas of Africa. It is also found in **slag** as a by-product of iron **smelting**.

Huge amounts (over one and one-half trillion tons) of manganese, along with smaller amounts of cobalt, nickel, and copper, are found in "nodules" (lumps) on the bottom of the Pacific Ocean. Mn nodules have also been found on the bottom of Lake Michigan. Proposals to mine the nodules have been suggested, but no large quantities have been recovered.

HISTORY

The Swedish chemist Carl Wilhelm Scheele (1742–1786) discovered many gases for which he did not receive credit, including oxygen, which he discovered two years before Joseph Priestley (1733–1804). However, because Scheele did not publish his findings in time, he did not get credit.

Scheele's work led to the discovery of manganese, for which John Gottlieb Gahn (1745–1818) received credit in 1774. Because Scheele did not recognize it or identify it as a new element and neglected to publish what he did know about this new element, he was not credited with the discovery of manganese.

COMMON USES

Manganese does not have many uses, but the utility it does have is important.

A major use is as an alloy to harden steel. At one time railroad rails were made of iron and had to be replaced about once a year because of the "softness" of unalloyed iron. Later it was discovered that with the addition of more than 20% of manganese to iron, it was possible to produce a superhard steel rail that would last over twenty years.

An important use is to remove oxygen and sulfur from molten iron during the process of making steel and alloys.

Manganese is also used in the production of aluminum and other metals to produce light, superhard tools that can withstand high temperatures, such as the tools used to cut metal on **lathes**.

It is used as a drying agent in paints and varnishes and as a bleaching agent for glass and oils.

Manganese dioxide (MnO_2) is the "black" content inside common dry cells (batteries) where it acts as an oxidizing agent.

A small amount of manganese (1.5 to 5 milligrams) is required in our diets for the normal development of tendons, bones, and some **enzymes**. Manganese is found in peas, beans, bran, nuts, coffee, and tea.

EXAMPLES OF COMMON COMPOUNDS

Several useful compounds are:

Potassium permanganate ($KMnO_4$)—A purplish compound used as an antiseptic and disinfectant to inhibit the growth of harmful skin microorganisms and bacteria. Before antibiotics were available, it was used as a treatment for trench mouth and impetigo.

Trench mouth (necrotizing ulcerative gingivitis), also called Vincent's infection, usually affects young adults and is considered a form of periodontal disease. If not treated, it can lead to the loss of gum tissue and eventually loss of teeth. Today there are more effective treatments for trench mouth than $KMnO_4$, so if your gums bleed frequently, see your dentist.

Impetigo is a skin infection that usually occurs in young children living in unsanitary conditions in warm climates. It is caused by either *Streptococcus* or *Staphylococcus* bacteria, or both together. It produces a tan scab that may disappear in about ten days, but should be treated with an antibiotic since it is contagious and can cause kidney damage.

Manganese carbonate ($MnCO_3$)—Used in the treatment of iron ore and as a **chemical reagent**.

Manganese gluconate [$Mn(C_6H_{11}O_7)$]—A food additive, vitamin, and dietary supplement (also, manganese glycerophosphate). Found in green leafy vegetables, legumes (peas and beans), and brewer's yeast.

Manganese chloride ($MnCl_2$)—Used in **pharmaceuticals**, in fertilizers, as a feed supplement, and as a dietary supplement.

Manganous oxide (MnO)—Used for textile printing, ceramics, paints, colored glass (green), bleaching, and fertilizers; as a dietary supplement; and as a **reagent** in analytical chemistry.

HAZARDS

The powder form of most manganese compounds, especially the oxides, is very toxic to plants, animals, and humans.

The dust or powder form of manganese is flammable. Water cannot be used to extinguish manganese fires. Manganese fires are extinguished by smothering with sand or dry chemicals.

❖ ❖ ❖

IRON (Period 4)
SYMBOL: Fe   **ATOMIC NUMBER:** 26   **GROUP:** 8

COMMON VALENCE: 2, 3, and 4 **ATOMIC WEIGHT:** 55.847 **NATURAL STATE:** Solid **COMMON ISOTOPES:** Four stable isotopes occur in nature: iron-54, iron-56, iron-57, and iron-58. There are 4 artificial radioisotopes produced in nuclear reactors. Two important ones are: iron-55, with a half-life of 2.9 years, and iron-59, with a half-life of 46 days, which is used as a **tracer** in medicine and metallurgy. **PROPERTIES:** Metallic iron is a silver-gray, tough, **magnetic**, **malleable** metal. It can be worked and made into many different shapes, such as rods, wires, sheets, ingots, pipes, framing, etc. Iron is very reactive and forms many compounds with other elements. There are two forms of the atom: ferrous iron (II—with a valence of 2) and ferric iron (III—with a valence of 3). Density: 7.87; melting point: 1536°C; boiling point: 3000°C.

CHARACTERISTICS

Iron is the only metal that can be tempered (hardened by heating, then quenching in water or oil). Iron is a good conductor of electricity and heat. It is easily magnetized but loses its magnetic properties at high temperatures.

Iron is the most important construction metal. It can be alloyed with many other metals to make a great variety of specialty products. Its most important alloy is steel.

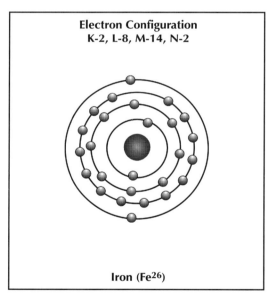

Electron Configuration
K-2, L-8, M-14, N-2

Iron (Fe^{26})

ABUNDANCE AND SOURCE

Iron is the fourth most abundant element, constituting about 5% of the Earth's crust. The core of the Earth is believed to be primarily molten iron and nickel.

Small amounts of iron alloyed with nickel and cobalt were found in meteorites by early humans. Even though small amounts of metallic iron are found in nature, early humans did not know how to extract iron from ores until well after they knew how to smelt gold, tin, and copper ores. From these metals they then developed bronze alloys, thus the Bronze Age. It is assumed that the reason for not using iron more widely is that early humans did not know how to make fires hot enough to smelt the iron from its ores.

There are several grades of iron ores: hematite (brown Fe_2O_3) and limonite (red Fe_2O_3). Other ores are pyrites, chromite, magnetite, siderite, and low-grade taconite. Magnetite (Fe_3O_4) is a magnetic iron mineral/ore found in South Africa, Sweden, and parts of the United States. The "lodestone," a form of magnetite, is a natural magnet that has polarity.

Iron ores are found in many countries. A major source in the United States is the Mesabi Range in Minnesota, which has produced over two billion tons of ore since it was first opened in 1884. Iron is also found in Alabama.

During the Revolutionary War, General George Washington made cannonballs and other iron products from the iron ore mined in the areas of Pine Grove Furnace and Cornwall, located in south-central Pennsylvania. Until recently, the Bethlehem Steel Company still used the iron ores from the old Cornwall deposits near Lebanon, Pennsylvania.

Iron is found in most of the universe, in most of the stars, and in our sun, and it probably exists on the other planets of our solar system.

HISTORY

The Iron Age followed the Bronze Age when people in Egypt, Asia, and Europe learned to use iron to make strong, sharp weapons and tools. As mentioned, early humans knew about iron but did not make much use of it until about 1000 B.C. This is about the time they learned how to **smelt** iron from its ore by first using charcoal, and later coke from coal, which made the fire hot enough to extract the iron from the impurities. The Iron Age began a new era for the advancement of wars, as well as benefiting mankind.

The knowledge of **metallurgy** in the smelting of iron ores, making of iron alloys, and working of iron into different shapes parallels the development of the industrial age. Iron is the least expensive of all metals to produce, and it is the most widely used metal in the world.

Wrought iron was used for beams to construct buildings and for weapons by the ancient Greeks and Romans. In addition, it was used for tools, vehicles, art, and coins. Its use spread to most other ancient countries, including India, China, and the Arab world.

COMMON USES

No one person can be given credit for the discovery of iron, but many people have contributed to our understanding of its chemistry and nature and of how to make practical use of this important element. A few examples are:

One of the major advances in the technology of iron smelting was the development of the Bessemer process by Henry Bessemer (1813–1898). In this process, compressed air or oxygen is forced through molten iron to oxidize (burn) out the carbon and other impurities, thus producing a more pure form of iron.

Powder metallurgy (**sintering**), i.e., the process whereby powdered iron and other metals are pressed together without heat to fit molded forms, is used to produce homogeneous (uniform throughout) metal parts.

Iron is used in the production of magnets of all shapes and sizes. Some supermagnets are mixtures of iron and other elements. These magnets are used in motors, atom smashers, and all types of electronic equipment, such as TVs and computers.

Iron is used to form wrought iron from "pig" iron, and when mixed with other metals to form alloys, iron takes on numerous characteristics for special uses, e.g., stainless steel.

Iron is an important element making up hemoglobin in the blood, which carries oxygen to the cells of our bodies. It is also very important as a trace element in the diet to assist with the oxidation of foods to produce energy. We need about 10 to 18 milligrams of iron each day, as a trace mineral. Iron is found in liver and meat products, eggs, shellfish, green leafy vegetables, peas, beans, and whole-grain cereals. Caution! Iron deficiency may cause anemia (low red blood cell count), weakness, fatigue, headaches, and shortness of breath. Excess iron in the diet can cause liver damage, but this is a rare condition.

EXAMPLES OF COMMON COMPOUNDS

Just a few examples are:

Iron oxides [Fe_2O_3; Fe_3O_4; $Fe(OH)_3$; and FeO], with individual colors, e.g., black, brown, metallic brown, red, and yellow—Used as catalysts, in glass and ceramic coloring, in **pigments**, in laundry blueing, and in steel making.

Ferrous sulfate ($FeSO_4$), also known as iron sulfate and iron vitriol—Used in the production of various chemicals, such as sulfur dioxide and sulfuric acid.

Pyrite (FeS_2), more commonly known as fool's gold—Used as an iron ore and to produce sulfur chemicals.

Ferrous chloride ($FeCl_2$)—Used in **pharmaceutical** preparations, for sewage treatment, and as a **mordant** (which fixes dyes so they will not run) in textiles.

HAZARDS

Iron dust and dust from most iron compounds are harmful if inhaled, and toxic if ingested. Iron dust and powder are also flammable and can explode in confined spaces. As mentioned, excessive iron in the diet may cause liver damage.

COBALT (Period 4)
SYMBOL: Co   ATOMIC NUMBER: 27   GROUP: 9

COMMON VALENCE: 2 and 3 **ATOMIC WEIGHT:** 58.933 **NATURAL STATE:** Solid **COMMON ISOTOPES:** The only stable isotope is cobalt-59. Several artificial radioisotopes are: cobalt-57, cobalt-58, and cobalt-60, which is the most important radioisotope. It is used for research in medicine and industry. A major use is for radiation treatment of some cancers. **PROPERTIES:** Cobalt is a shiny, steel-gray, hard, brittle metal that is not very **malleable**. Density: 8.9; melting point: 1493°C; boiling point: 3100°C.

CHARACTERISTICS

Cobalt has chemical properties closely related to those of iron and nickel in both its metallic and combined states.

It is attacked by most acids and will corrode in air. When alloyed with iron, it makes excellent magnets.

ABUNDANCE AND SOURCE

Cobalt is the thirty-second most abundant element on Earth. It is not found in a free state, but is widely distributed in **igneous rocks**, although not in high concentrations. It occurs in most soils of the Earth, in all animals and plants, in both fresh and seawater, in the stars, and in meteorites.

Cobalt is separated from the following ores: cobaltite, linnaetite, chloanthite, and smaltite. Traces are also found in silver, copper, nickel, iron, zinc, and manganese ores.

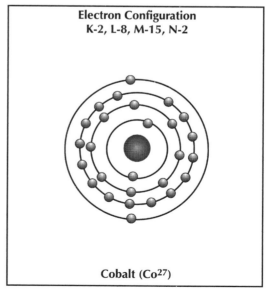

Electron Configuration
K-2, L-8, M-15, N-2

Cobalt (Co^{27})

The major ores are found in Canada and parts of Africa.

HISTORY

In 1737 Georg Brandt (1694–1768) was investigating a bluish mineral thought at one time to be copper ore. The early **smelters** could not get copper from this particular ore. They thought that the gnomes (*kobolds*, which means "goblins" in German) of the mountains where they mined the ore had put an evil spell on it to prevent the ore from producing copper. Of some interest is that the element nickel had a similar problem with superstitious miners. Brandt isolated cobalt as a new element and named it after the earth spirit (good spirits).

Cobalt was the first metal to be discovered that was not known to the ancient alchemists. Brandt is considered one of the first chemists of the new era no longer under the influence of the old alchemists.

COMMON USES

Cobalt has many uses.

A major use is in alloys with iron and chromium to produce high-speed, high-temperature tools and jet engine parts.

Cobalt alloys also make high-grade soft and hard permanent magnets. It is also used for dental and medical equipment.

Cobalt is used in paints, glass and ceramic manufacturing, inks, and as a **catalyst** in chemical reactions.

The radioisotope cobalt-60 is used to test welds and casts for flaws, for irradiation of food crops, as a portable source of ionizing **radiation**, and for medical and other research purposes.

It is a natural component of vitamin B_{12} and, thus, is important as a trace element in our diet.

EXAMPLES OF COMMON COMPOUNDS

Cobalt blue [$Co(AlO_2)_2$], also known as cobalt ultramarine or azure blue; a compound of aluminum oxide and cobalt—Used as a **pigment** that mixes well with both oil and water. As a cosmetic, it is used as an eye shadow and in grease paint.

Cobaltic oxide (Co_2O_3), also known as cobalt oxide or cobalt black; dark gray to black—Used in pigments, ceramic glazes, and **semiconductors**.

Cobalt chloride ($CoCl_2$)—Used to manufacture vitamin B_{12}, even though the compound itself can cause damage to the blood cells. Also used as a dye **mordant** (to fix the dye to the cloth), to make solid lubricants, as a **reagent** chemical in chemical reactions, and as an additive in fertilizers.

HAZARDS

Cobalt dust and fine powder are both flammable and toxic if inhaled.

Cobalt is found in most natural foods and is toxic to humans only if ingested in the digestive system or injected in the blood in large amounts. Even though the human system will rapidly eliminate excessive cobalt (and vitamin B_{12}) in the urine, large intakes of it can cause illness and even death.

Some years ago a cobalt additive was used by some beer makers to maintain a foam head on their beer. Those who imbibed excessively developed what was known as "beer drinker's syndrome," which caused some deaths due to enlarged, flabby hearts.

Cobaltous chromate ($CoCrO_4$), as well as some other cobalt compounds, is **carcinogenic**.

❖ ❖ ❖

NICKEL (Period 4)
SYMBOL: Ni   **ATOMIC NUMBER:** 28   **GROUP:** 10

COMMON VALENCE: 2 and 4 **ATOMIC WEIGHT:** 58.70 **NATURAL STATE:** Solid **COMMON ISOTOPES:** Five stable natural isotopes exist: nickel-58, nickel-60, nickel-61, nickel-62, and nickel-64. Seven radioactive isotopes exist: nickel-56, nickel-57, nickel-59, nickel-63, nickel-65, nickel-66, and nickel-67. **PROPERTIES:** Nickel ores are yellowish, while the metal itself is a silver-white color. Nickel is hard but **malleable** and **ductile** and can be worked hot or cold to fabricate many items. Density: 8.908; melting point: 1455°C; boiling point: 2900°C.

CHARACTERISTICS
Nickel is located in Group 10 (VIII) and is the third element in the special Group of the first series of transition metals. It is similar to Fe and Co.

Some **acids** will attack nickel, but it offers good protection from corrosion from air and seawater. Nickel makes excellent alloys with several other metals.

ABUNDANCE AND SOURCE
Nickel is the twenty-third most abundant element on

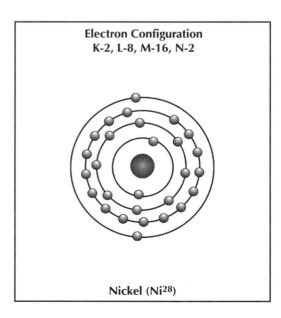

Electron Configuration
K-2, L-8, M-16, N-2

Nickel (Ni[28])

Earth. It is somewhat plentiful, but scattered. It makes up about one-hundredth of 1% of **igneous rocks**. Pure nickel metal is also found in meteorites.

There are two main types of nickel ore. One is the iron/nickel sulfide ores called pentlandite and pyrrhotite. The other ore is an oxide of nickel called garnierite.

Nickel ores are found in Ontario, Canada, Cuba, the Dominican Republic, and Norway.

Traces of nickel exist in soils, coal, plants, and animals.

HISTORY

Early miners had trouble **smelting** copper from some of the ores because they kept getting other metals (see the reference to cobalt). This caused them so much trouble that they named the element *kupfernickel*, or "Old Nick's copper," which meant the *devil's copper*.

In 1751 Axel Fredrick Cronstedt (1722–1765) used some of the techniques he learned from his teacher, Georg Brandt, to separate a "new" metal that was neither copper nor cobalt from the ore. He named it *nickel* (after "Old Nick") since that was the name the early miners called the ore.

COMMON USES

The most important use of nickel is as an alloy metal with iron and steel to make stainless steel, which contains from 5% to 8% nickel. It is also alloyed with copper to make Monel metal, which was used to make rust-resistant sinks, etc., before stainless steel became economical.

Nickel/copper alloy was used to make coins for thousands of years. Only recently did the United States start to use nickel alloyed with copper, zinc, and other metals to replace silver in several coins, such as the nickel, dime, quarter, and half and silver dollars. There are other useful alloys of nickel with aluminum, chromium, and brass. Nickel alloys are also used to make bullet-proof and armor plate.

Nickel is used in batteries and as **electrodes** in **fuel cells** (see the reference for hydrogen). It is also used in the **hydrogenation** of liquid vegetable oils to convert them to solids.

EXAMPLES OF COMMON COMPOUNDS

Nickel chloride ($NiCl_2$)—Used for **electroplating** nickel on other metals and as a chemical **reagent** in laboratories.

Nickel oxide (NiO)—Used to make porcelain painting and fuel cell electrodes.

Nickel-silver, not a compound, but an alloy of nickel with copper and zinc—Used in etching and in the process to plate silver and chromium.

Nickel sulfate (NiSO₄), known also as nickel salts or blue salt—Used in nickel plating, coatings, and ceramics. It is also useful as a **mordant** (color fixer) for dyeing textiles.

Due to its excellent heat conductivity and resistance to corrosion, nickel is an excellent metal for cookware.

HAZARDS

Nickel dust and powder are flammable. Most nickel compounds, particularly the salts, are toxic.

Although nickel is not easily absorbed in the digestive system, it can cause a toxic reaction and has been shown to be **carcinogenic** (cancer causing) in high concentrations in the body. Nickel workers can receive severe skin rashes and lung cancer from exposure to nickel dust and vapors.

Nickel is stored in the brain, spinal cord, lungs, and heart. It can cause coughs, shortness of breath, dizziness, nausea, vomiting, and general weakness.

❖ ❖ ❖

COPPER (Period 4)
SYMBOL: Cu   **ATOMIC NUMBER:** 29   **GROUP:** 11

COMMON VALENCE: 1 and 2 **ATOMIC WEIGHT:** 63.546 **NATURAL STATE:** Solid **COMMON ISOTOPES:** Two stable isotopes exist: copper-63 and copper-65, plus 9 radioactive isotopes of copper. **PROPERTIES:** Native copper has a distinctive brownish/reddish color. Its two types of ions are: cuprous Cu(I), which oxidizes easily, and cupric Cu(II), whose compounds are stable. Copper is malleable and ductile. It forms a greenish film over its surface when exposed to moist air, which protects the metal from further corrosion. Density: 8.896; melting point: 1083°C; boiling point: 2595°C.

CHARACTERISTICS

Copper is an excellent **conductor** of heat and electricity.

It can be worked into various forms such as pipes, wire, rods, sheets, and coins.

Copper will dissolve in acids, but is resistant to atmospheric corrosion.

It is more resistant to corrosion than iron. Copper is a necessary trace element in the diet of plants and animals.

ABUNDANCE AND SOURCE

Copper is the twenty-sixth most common element on Earth. As a pure metal it is rare, but it is found in many types of ores in most areas of the world—mostly in quantities not cost-effective to mine.

Copper is found in two types of ores: (1) sulfide ores, such as covellite, chalcopyrite, bormite, chalcocile, and enargite; (2) oxidized ores, such as tenorite, malachite, azurite, cuprite, chrysocolla, and brochanite.

Copper ores are found in Michigan, Arizona, Utah, Montana, New Mexico, Nevada, Tennessee, Chile, Canada, Russia, and Africa.

High-grade native ores, once found in the United States, have long been exhausted, but many low-grade ores still exist in great quantities.

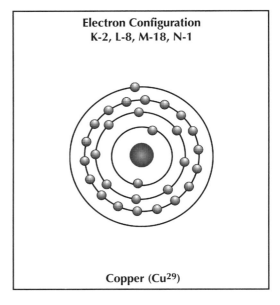

Electron Configuration
K-2, L-8, M-18, N-1

Copper (Cu²⁹)

HISTORY

Nuggets of high-grade copper were found by early humans sometime around 5000 B.C. Even though it is believed that copper was one of the first metals used by humans, they did not find much use for pure copper. People learned to melt copper ores in hot fires where the carbon (charcoal) from the wood combined with impurities to form CO_2 and high-grade metallic copper. Once smelted, the metal could be shaped for greater use.

Pure copper was too soft for much use. Probably by accident, early humans learned to **alloy** copper, first with arsenic, which proved to be poisonous, and later with tin ore, which produced an alloy that would keep a sharp edge, i.e., bronze. Thus, copper is thought to be one of the first metals to be found, **smelted**, and **alloyed** by early humans, perhaps as long ago as 4000 B.C.

Copper's name comes from the Latin word *cuprum*, or *cyprium*, which is related to the name "Cyprus," the island where it was found by the Romans in early times.

COMMON USES

There are many uses for copper. It is a very versatile metal that has increased in importance to technology and society in the last century.

It is one of the three inert "coin" metals. Along with gold and silver, it has been used to make coins for many centuries.

Some major uses are: electrical wiring and components, roofing, plumbing, alloys (brass, bronze, and Monel metal), **electroplating** (using an electric charge to deposit a thin coating of a metal on another object), cooking utensils, insecticides, marine paints, cosmetics, and wood preservatives.

EXAMPLES OF COMMON COMPOUNDS

When copper, with a valence of 1, forms compounds, the compounds are known as "**cuprous** compounds." When the valence is 2, they are known as "**cupric** compounds," which are more stable than the cuprous compounds.

Copper chloride ($CuCl_2$), also known as cupric chloride since it has a +2 valence—Used as a catalyst, in **metallurgy**, in the textile industry, as a disinfectant, for green-colored fireworks, in wood preserving, for red-colored glass and ceramics, plus many other uses.

Many of the copper compounds have similar uses, as in fungicides (to kill fungi and mold plants), **insecticides**, **pigments**, electroplating, and preservatives.

HAZARDS

Most of the compounds of copper are toxic by either ingestion or inhalation.

The dust and powder of metallic copper and some compounds of copper are flammable, or explosive when ignited in contained areas.

Plants and animals require a trace of copper in their diet, but excessive copper is toxic.

ZINC (Period 4)
SYMBOL: Zn   **ATOMIC NUMBER:** 30   **GROUP:** 12

COMMON VALENCE: 2   **ATOMIC WEIGHT:** 65.38   **NATURAL STATE:** Solid
COMMON ISOTOPES: Zinc is not found as a native metal. There are 5 stable isotopes: zinc-64, zinc-66, zinc-67, zinc-68, and zinc-70. A major radioisotope is zinc-65, which is produced in nuclear reactors and **cyclotrons**.
PROPERTIES: Zinc is a whitish metal with a bluish hue. Zinc is an **electropositive** metal (will give up its outer two electrons in the N shell to nonmetals). Zinc foil will ignite in moist air, and zinc shavings and powder react violently with acids. Density: 7.14; melting point: 419°C; boiling point: 907°C.

Note: Zinc is not always included as one of the metals in the first series of the transition elements, but it is included in Group 12.

CHARACTERISTICS

Zinc is **malleable** and produces many useful alloys.

It is not magnetic but does resist **corrosion**, thus its use as a coating for other metals.

Zinc is an important and essential element for proper nutrition, but excessive zinc in the diet may cause zinc **intoxication**.

ABUNDANCE AND SOURCE

Zinc is the twenty-fourth most abundant element and is not found in its pure form in nature. It is found in several minerals and ores.

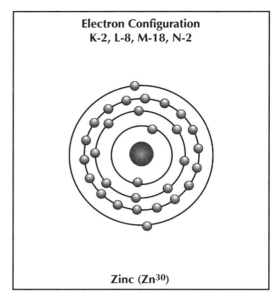

Electron Configuration
K-2, L-8, M-18, N-2

Zinc (Zn³⁰)

It is recovered from ores and minerals, such as willemite, hydrozincite, smithsonite, wurtzite, zincite, and Franklinite. These ores are found in Canada, Mexico, Australia, Belgium and the United States. Two of the ores, Franklinite and willemite, are valuable grade zinc ores mined in New Jersey.

HISTORY

No one person is credited with the discovery of zinc. It may have been isolated and used as a metal in India, but we do not know for sure. Its ores were known to early humans, but the metal was mixed with other metals, such as lead, arsenic, antimony, and bismuth.

In the sixteenth century scientists developed assaying equipment that enabled them to separate the different metals that were found in ores. Theophrastus Bombastus von Hohenheim (1493–1541), better known as Paracelsus, was given credit for the discovery of zinc.

In 1659 Johann Rudolf Glauber (1604–1668) recognized that different elements produced different-colored flames when heated. He used the "glass bead" analysis technique to identify zinc as well as many other metals that were often confused with each other.

Glauber also produced many medicines of his day, including his *sal mirabile*, or "wonderful salt," which he advertised as a cure-all because of its laxative properties. Today it is called Glauber salt (sodium sulfate).

COMMON USES

The main uses of zinc are for **electroplating** and in special alloys with other metals.

There are several methods of coating zinc on iron or steel. One is to immerse the iron in molten zinc to **galvanize** the iron. This is called the hot-dip process that deposits a thin coating of crystallized zinc onto the surface of the iron. The resistance of the galvanized iron item to corrosion is dependent on the thickness of the zinc coating.

Another method is to electroplate the iron with zinc, using an electric current to deposit a thin coating of zinc on the iron.

And a third is to "metallize" the iron by spraying it with very hot zinc powder.

Zinc is used with copper and other metals to produce alloys of brass, bronze, and special die-casting alloys, e.g., copper, aluminum, nickel, and titanium.

It is used in batteries, in **fungicides** (to kill fungi and molds), for roofing, and in wrapping wires for protection.

Zinc is an important trace element required for all healthy plants and animals. Zinc is found in proteins, such as meats, fish, eggs, and milk. About 10 to 15 milligrams is required per day. It may be taken as a dietary supplement. Zinc helps the blood in our bodies move the waste gas—carbon dioxide—to the lungs, and helps prevent macular degeneration (loss of vision).

EXAMPLES OF COMMON COMPOUNDS

There are hundreds of uses for zinc. Following are examples:

Zinc acetate [$Zn(C_2H_3O_2)_2$]—Used as a **mordant** for dyeing cloth, as a wood preservative, as a laboratory agent, and as a diet supplement if one does not get enough zinc in the diet.

Zinc chloride ($ZnCl_2$)—Used as an organic catalyst, as a dehydrating and condensing agent, in electroplating for an undercoating of other metals, as an antiseptic, as a component of some deodorants, and as an astringent. It is **deliquescent**, which makes it an excellent drying agent.

Zinc oxide (ZnO), known as zinc white when used as an oil paint by artists—Added to paint as a mold inhibitor; used for cosmetics, seed treatments, pigments, and as a dietary supplement.

Zinc sulfate ($ZnSO_4$)—Used in the textile industry and to make rayon, as a wood preservative, an analytical **reagent**, and as a diet supplement. It can also be used to stop bleeding.

HAZARDS

Zinc dust and powder are very explosive. When placed in acid or a strong alkaline solution, zinc metal shavings will produce hydrogen gas which may explode.

A deficiency of zinc in humans will retard growth, both physical and mental, and contribute to anemia. It is present in many foods, particularly proteins. A balanced diet provides an adequate amount of zinc. Not more than 50 milligrams per day of dietary zinc supplement should be taken, since high levels of zinc are toxic. Our bodies normally contain about 2 grams of zinc. A lack of zinc in the diet may cause a lack of taste and delay growth and cause retardation in children.

Zinc intoxication can occur both from inhaling zinc fumes and particles, mainly in industrial processes, or from orally ingesting too much zinc dietary supplement (vitamins). Zinc intoxication can cause stomach pains, vomiting, and bleeding. Excessive zinc can cause premature birth in pregnant women.

Second Series
(PERIOD 5; GROUPS 3 to 12)

YTTRIUM (Period 5)
SYMBOL: Y   **ATOMIC NUMBER:** 39   **GROUP:** 3

COMMON VALENCE: 3 **ATOMIC WEIGHT:** 88.906 **NATURAL STATE:** Solid **COMMON ISOTOPES:** Yttrium-89 is the only stable isotope. **PROPERTIES:** Yttrium is always found with the rare-Earth elements, and in some ways it resembles them. Although it is sometimes classified as a rare-Earth element, it is listed in the Periodic Table as the first element in the second row (Period 5) of the transition metals. Density: 4.47; melting point: 1500°C; boiling point: 2927°C.

CHARACTERISTICS

Yttrium is a dark-gray metallic element that is soluble in both dilute acids and basic solutions.

Yttrium was extremely difficult to separate and analyze as a new element because most of the rare-earth elements have very similar characteristics.

ABUNDANCE AND SOURCE

Yttrium is the twenty-ninth most abundant element found on the Earth. It is found as an oxide in the same deposits of monazite ores as are lanthanum and the rare-Earth elements.

The name "rare-Earth elements," including yttrium, does not mean that they are scarce. It was just difficult to identify them.

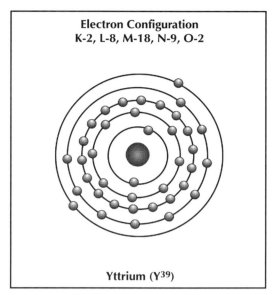

Electron Configuration
K-2, L-8, M-18, N-9, O-2

Yttrium (Y³⁹)

HISTORY

In 1787 a Swedish amateur mineralogist, Carl Axel Arrhenius (1757–1824) discovered a rock he named *ytterite* after a mine located near the town of Ytterby in Sweden.

In 1794 Johan Gadolin (1760–1852) analyzed the rock and found that it contained a new "earth." It was then named *gadolinite*. This black earth was further examined and identified by other chemists of the day including Carl Gustav Mosander. Gadolin was given the credit for discovering yttrium, even though some references give credit for discovery to Friedrich Wohler (1800–1882) who isolated it in 1828. It was named *yttria*, and then finally *yttrium*.

COMMON USES

There are few uses for yttrium, but they are important ones.

Since it emits a very bright red color when excited by electrons, it is used as a coating on TV screens, to produce the reds in television pictures.

It is used as an alloy in producing special types of iron and steel, and it has some uses in nuclear technology.

It is also used in the production of several types of **semiconductors**.

EXAMPLES OF COMMON COMPOUNDS

Yttrium arsenide (YAs)—Used for the production of high-grade semiconductors. Since it is extremely toxic, special handling and facilities are required.

Yttrium oxide (Y_2O_3)—Used for television tubes to produce red colors.

Yttrium chloride (YCl₃)—Decomposes at a relatively low temperature (100°C) and is useful in laboratories as a chemical **reagent** that identifies other substances.

HAZARDS

As a powder or fine particles, yttrium is flammable in air. Some of its compounds, particularly those used in the semiconductor, electrical, and other industries, can be very toxic if ingested or inhaled.

❖   ❖   ❖

ZIRCONIUM (Period 5)
SYMBOL: Zr   ATOMIC NUMBER: 40   GROUP: 4

COMMON VALENCE: 2, 3, and 4 **ATOMIC WEIGHT:** 91.22 **NATURAL STATE:** Solid **COMMON ISOTOPES:** Zirconium has 5 stable isotopes: zirconium-90, zirconium-91, zirconium-92, zirconium-94, and zirconium-96. There is 1 important radioisotope: zirconium-95, with a half-life of 63 days. It emits both beta and gamma radiation. **PROPERTIES:** Zirconium is a crystal-like hard, strong, **ductile**, and **malleable** metal with lustrous grayish color. It is a very reactive metal, so it is found only in compounds with other elements (mostly oxygen). Zr^{40} is very difficult to distinguish from the element hafnium (Hf^{72}), which is located just below Zr in the Periodic Table. Density: 6.5; melting point: 1850°C; boiling point: 4377°C.

CHARACTERISTICS

Zirconium is insoluble in water and cold acids.

Although it is a reactive metal, it resists corrosion due to rapid surface oxidation, which produces a protective film that makes it less reactive with other elements.

It also has a valence of 4 when combined with one of the halogens from Group 17 (VIIA). ZrO_2 is its most common compound.

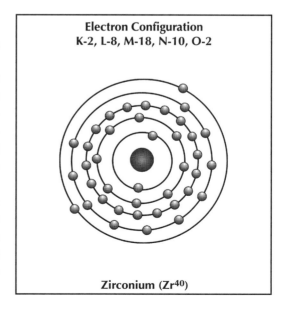

Electron Configuration
K-2, L-8, M-18, N-10, O-2

Zirconium (Zr^{40})

ABUNDANCE AND SOURCE

Zirconium is not a rare element. It is found over most of Earth's crust and is the eighteenth most abundant element, but is not found as a free metal in nature.

It is found in the ores baddeleyite (also known as zirconia) and in the oxides of zircons, elpidite, and eudialyte.

HISTORY

In 1789 Martin Heinrich Klaproth (1743–1817) (who also discovered uranium) identified the element zirconium, which was found in *zircon* gem stones, known since ancient times and from which it gets its name. The metal itself was not separated from its ores until 1824 by Jöns Jakob Berzelius (1779–1848).

Zircon is a natural zirconium silicate oxide that accounts for about 75% of the zirconium that is used.

COMMON USES

The most important use is to wrap thin sheets of zirconium around the fuel rods and other construction material in nuclear power plants. For this purpose it must be very pure and not mixed with other elements such as its "cousin" hafnium.

Zirconium is an important **abrasive** used in industry. It makes excellent grinding wheels and special sandpapers.

It is used in ceramic glazes, in enamels, and for linings in furnaces and molds. It resists corrosion at high temperatures, which makes it ideal for laboratory ware.

EXAMPLES OF COMMON COMPOUNDS

Zirconium oxide (ZrO_2), the most common compound; usually found in nature as oxide compounds and then refined—It has many uses, including making heat-resistant fabrics and high-temperature **electrodes** and tools, treating skin diseases and, of course, producing the elemental metal zirconium.

Zirconium carbide (ZrC)—Used for light bulb filaments, metal **cladding**, abrasives, and refractory furnace linings.

Zirconium sulfate [$Zr(SO_4)_2$]—Used to make high-temperature lubricants, for tanning white leather, as a **chemical reagent,** and as a catalyst in chemical laboratories.

Zirconium-95, the radioactive isotope of zirconium—Used to trace the flow of fluids in oil pipelines and as a catalyst in **cracking** plants that produce petroleum products from crude oil.

Zircons, sometimes referred to as "cubic zircons," are clear, colorless semiprecious gemstones. Chemically, zircons are natural zirconium silicate

($ZrSiO_4$), which are polished to look like fake diamonds. Zirconium silicate is also found in brown, red, and gray.

As a compound of zirconium, aluminum, and chlorine, it is used in roll-on deodorants to prevent perspiration.

HAZARDS

There is not much agreement about how dangerous the elemental form of zirconium is. Some say that the metal and gem forms are harmless, but there is some evidence that the vapors and powder forms may be **carcinogenic**. Some people seem to be allergic to several zirconium compounds.

Zirconium fine powder and dust are explosive, especially in the presence of oxidizers (nonmetals that "grab" or share electrons from the zirconium atoms).

Some compounds of zirconium have proven to be toxic to the skin or lungs if inhaled. Zirconium metal (powder, foil, small pieces, etc.) should be stored under water to prevent oxidation.

❖ ❖ ❖

NIOBIUM (Period 5)
SYMBOL: Nb   **ATOMIC NUMBER:** 41   **GROUP:** 5

COMMON VALENCE: 2, 3, 4, and 5   **ATOMIC WEIGHT:** 92.906   **NATURAL STATE:** Solid   **COMMON ISOTOPES:** No stable isotopes are known. **PROPERTIES:** Niobium is a grayish-silver soft metal that does not **tarnish** (oxidize) at room temperature. It is closely related to tantalum, which is located just below it in Group 5 of the Periodic Table. Density: 8.57; melting point: 2468°C; boiling point: 5127°C.

CHARACTERISTICS

Niobium is a **ductile** and **malleable** metal that can be drawn into wires or hammered into shapes.

It is noncorrosive at normal temperatures, but will react with oxygen and halogens (Group 17) at high temperatures.

ABUNDANCE AND SOURCE

Niobium is the thirty-third most abundant element on Earth, and it is considered rare. Niobium does not exist as a free metal in nature, but rather it is found in two ores, columbite and pyrochlore. These ores are found in Canada and Brazil. Niobium and tantalum are also products from tin mines in Malaysia and Nigeria. Niobium is a chemical "cousin" of tantalum and must be purified by separating it by a process known as **fractional crystal-**

lization (separation is accomplished due to the different rates at which some elements crystallize), or by dissolving it in special **solvents**.

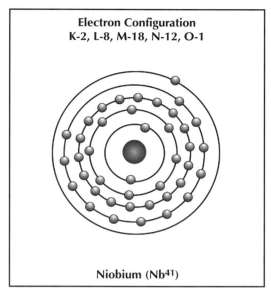

Electron Configuration
K-2, L-8, M-18, N-12, O-1

Niobium (Nb⁴¹)

HISTORY

Charles Hatchett (1765–1847) examined a sample of a mineral sent from the United States to the British Museum. He identified what he considered a new element and called it *columbium*, for Columbia, another name for the American continent.

By the time the substance was rediscovered and confirmed as a new element by Heinrich Rose (1847–1920), the scientific world had named it *niobium* after Niobe, the daughter of Tantalus in Greek mythology.

Many metallurgists in the United States still refer to it by the name Hatchett first gave it—*columbium*. Hatchett was given credit for its discovery.

COMMON USES

Niobium has special **cryogenic** properties. It can withstand very cold temperatures, which improves its ability to conduct electricity. This characteristic makes it an excellent metal for low-temperature electrical **superconductors**.

It is also used to produce special stainless steel alloys, to make high-temperature magnets, as special metals for rockets and missiles, and for high- and low-temperature-resistant ceramics.

EXAMPLES OF COMMON COMPOUNDS

Niobium carbide (NbC)—Used to make hard-tipped tools and special steels, and to coat graphite in nuclear reactors.

Niobium chloride ($NbCl_5$)—Used to produce pure niobium.

Niobium silicide ($NbSi_2$)—Used to line high-temperature refractory furnaces.

Niobium-uranium alloy—Makes excellent fuel rods for **nuclear reactors** because it is very hard with a high **tensile** strength. It won't pull apart and break down in the reactor.

As a metal, niobium is a superconductor of electricity—Used to make powerful magnets that are experimentally being tested to "drive" superfast forms of ground and sea transportation.

HAZARDS

Niobium is not considered very toxic, since it does not react with many substances at room temperature. But, as with most metals, the dust, powder, or vapors of niobium should be avoided. They can cause severe irritation and possibly cancer if ingested or inhaled.

MOLYBDENUM (Period 5)
SYMBOL: Mo   **ATOMIC NUMBER:** 42   **GROUP:** 6

COMMON VALENCE: 2, 3, 4, 5, and 6 **ATOMIC WEIGHT:** 95.94
NATURAL STATE: Solid **COMMON ISOTOPES:** Molybdenum has 7 stable isotopes. **PROPERTIES:** Molybdenum is a hard but **ductile** and **malleable** silver-white metal with a high melting point. It does not occur free in nature. It can be extracted from its ores as either a black powder or grayish metal. Density: 10.2; melting point: 2610°C; boiling point: 5560°C.

CHARACTERISTICS

Molybdenum oxidizes at high temperatures, but not at room temperatures. It is insoluble in acid and **hydroxides** at room temperatures.

It can be worked into wire, rods, and sheets, as well as formed into **ingots** from powder and crystals.

ABUNDANCE AND SOURCE

Molybdenum is the fifty-fourth most abundant element. It is found throughout the world, but not in its pure form. There are not many concentrated sites worth mining. It is recovered from some copper-mining operations, as well as from molybdenum mines.

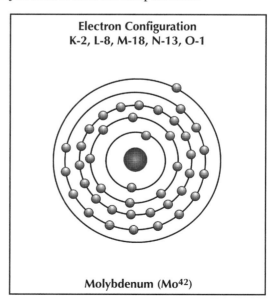

Electron Configuration
K-2, L-8, M-18, N-13, O-1

Molybdenum (Mo⁴²)

The chief ore, molybdenite (MoS_2), is found in China, Mexico, Australia, England, Norway, Sweden, and parts of the United States.

Molybdenite ore is very similar to **graphite**, and they have been mistaken for each other in the past.

HISTORY

Peter Jacob Hjelm (1746–1813) is given credit for discovering molybdenum in 1781 despite the fact that his paper was not published until 1890. He followed the advice of Carl Wilhelm Scheele (1742–1786), who isolated and identified molybdenum, but incorrectly thought it was an element related to lead.

Although some reference works do give Scheele credit, most do not credit him for the discovery of molybdenum, or for other elements he "discovered," such as oxygen and manganese.

Scheele did not receive credit for discovering oxygen two years before Joseph Priestley (1733–1804) announced his discovery and was given the credit. Scheele's publisher was negligent in getting his work published in time.

There may be a lesson in this story for all young scientists—keep complete and accurate records of all your lab work and observations, and when you are sure of your experimental results—PUBLISH!

COMMON USES

The most important uses are to make special alloys of steel, iron, and other metals of great strength, and for high-temperature tools. Molybdenum is also used as a protective coating of metals so they can withstand high temperatures.

Molybdenum is used for making special parts for missiles, rockets, and aircraft.

It has use as a pigment for paints and printing ink, and as a high-temperature lubricant for spacecraft and automobiles.

Molybdenum has a high resistance to electricity, which makes it ideal as a filament wire for electric heaters.

EXAMPLES OF COMMON COMPOUNDS

Molybdenum boride (Mo_2B)—Used to **braze** (weld) special metals and for noncorrosive electrical connectors and switches. Used for cutting tools and noncorrosive, abrasion-resistant parts for machinery.

Molybdenum disulfide (MoS_2)—Used to make special greases and lubricants that withstand high temperatures.

Molybdenum pentachloride ($MoCl_5$)—Used as a **brazing** and **soldering flux** and to make fire-retardant resins.

HAZARDS

The powder and dust forms of molybdenum are flammable. The fumes from some of the compounds should not be inhaled or ingested.

Only some of the compounds of molybdenum are toxic, particularly in mist or powder form.

Small traces of molybdenum are essential for plant and animal nutrition.

❖   ❖   ❖

TECHNETIUM (Period 5)
SYMBOL: Tc   ATOMIC NUMBER: 43   GROUP: 7

COMMON VALENCE: 4, 5, 6, and 7   **ATOMIC WEIGHT:** 98.906   **NATURAL STATE:** Solid   **COMMON ISOTOPES:** There are no stable forms of technetium; all are radioactive. Technetium was the first element to be artificially produced in a **cyclotron**.

It is now produced as a **fission** product in nuclear reactors. Two of its radioisotopes have very short half-lives and are unstable, but technetium-99 has a **half-life** of over 105 years. **PROPERTIES:** Technetium was the first element, not found on Earth, to be artificially produced by bombarding molybdenum with **deuterons**. It is silver-gray in color.

Chemically, technetium resembles the element rhenium (Re[75]), which is just below it in Group 7 of the Periodic Table. At one time it was also thought to be similar to manganese, and for a short time was referred to as **eka**-manganese. Density: 11.5; melting point: 2200°C; boiling point: 4877°C.

CHARACTERISTICS

Technetium's chemistry is very similar to that of rhenium, and it forms similar compounds.

It exists in very minute amounts along with some other elements as unstable isotopes.

It is the lightest element to have no stable isotopes found in nature. Technetium has excellent **superconductivity** properties.

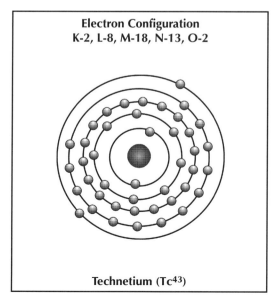

Electron Configuration
K-2, L-8, M-18, N-13, O-2

Technetium (Tc[43])

ABUNDANCE AND SOURCE

Technetium is the seventy-sixth most abundant element and only traces of it are found on Earth in pure form. Although it is really not found on Earth, it has been identified in the light from some stars by using a **spectroscope**.

The man-made technetium radioisotopes are artificially produced in cyclotrons and nuclear reactors.

The long-half-life radioisotope technetium-99 exists for periods long enough for study and use.

HISTORY

Several chemists speculated about technetium, but were never able to find any of it.

In 1937 Emilio Gino Segre (1905–1989) and Carlo Perrier (birth date unknown) postulated that if they could not find technetium in the Earth's crust, perhaps they could make it.

Previously, Enrico Fermi (1901–1954) had changed one element to another by bombarding the nuclei with deuterons [atomic nuclei of heavy hydrogen (H^2) which have 1 proton and 1 neutron].

Segre and Perrier knew about Fermi's work and decided to bombard element 42 (molybdenum) with these heavy hydrogen nuclei in a cyclotron (particle accelerator) and convert it to element 43 (technetium). Their technique worked. And they were jointly given the privilege of naming the first artificially produced element, just as other scientists were for naming naturally occurring elements they discovered. In 1939 they named it *technetium* from the Greek word for "artificial."

Of some interest was that Mendeleyev (developer of the Periodic Table) had the insight to figure out, based on their position in his Table, some of the properties of elements not yet discovered.

He called these *eka* elements, which means "first" in Sanskrit. Once discovered, they were given names by the discoverer. Some examples of eka elements predicted by their placement on the Periodic Table (some were accurate—others just close) were eka-aluminum, eka-gallium, eka-boron, eka-scandium, eka-silicone, and eka-manganese.

COMMON USES

Technetium is a rare element, and there are not many uses for it. One important use is to produce strong magnets that can be supercooled to increase their efficiency. These supercooled magnets have practical uses in some industries, such as imaging technology, electrical generation and propulsion, and for experimental research where supercooled magnets are being tried out as a means to propel high-speed, lightweight trains.

Technetium is used as a **tracer** in producing various metals and in the production of corrosion-resistant alloys. Technetium-99 is a source of radiation used in nuclear medicine.

EXAMPLES OF COMMON COMPOUNDS
Only a few compounds have been made. They are all oxides of technetium: TcO_2, Tc_2O_7, and NH_4TcO_4.

HAZARDS
The hazards of technetium are the same as for all radioactive elements. Excessive exposure to radiation can cause many kinds of tissue damage— from sunburn to radiation poisoning and death.

RUTHENIUM (Period 5)
SYMBOL: Ru   **ATOMIC NUMBER:** 44   **GROUP:** 8

COMMON VALENCE: 3, 4, 5, 6, and 8 **ATOMIC WEIGHT:** 101.07 **NATURAL STATE:** Solid **COMMON ISOTOPES:** Seven stable isotopes are known. **PROPERTIES:** Ruthenium is a rare, hard, silvery-white metallic element located in Group 8, just above osmium. Ruthenium has many chemical and physical properties similar to osmium. Density: 12.41; melting point: 2310°C; boiling point: 4100°C.

CHARACTERISTICS
Ruthenium belongs to the "platinum group," which includes six elements with similar chemical characteristics that are located in the middle of the transition elements (Groups 8, 9, and 10). The elements of the platinum group are: ruthenium, rhodium, palladium, osmium, iridium, and platinum. Ruthenium is not attacked by acids, but strong alkalies will corrode it.

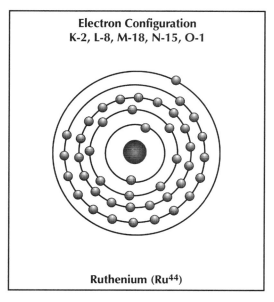

**Electron Configuration
K-2, L-8, M-18, N-15, O-1**

Ruthenium (Ru⁴⁴)

ABUNDANCE AND SOURCE

Ruthenium is the seventy-fourth most abundant element on Earth. It is considered a rare element, and small amounts of it are found in platinum deposits.

It is found in South America and the Ural Mountains of Russia. There are minor deposits in the United States and Canada.

Radioisotopes of ruthenium are produced in **nuclear reactors**.

HISTORY

In 1748 Antonio de Ulloa (1716–1795), a Spanish scientist and explorer, reported finding a special metal in South America. It was silvery-gray, more dense than gold, but it did not have the attractive luster of gold or silver. He did not know that he had located a new element along with the platinum metal.

It was not until 1844 that Karl Karlovich Klaus (1796–1864) separated ruthenium from platinum and correctly identified its properties. He is given credit for its discovery and he named it *ruthenian* after Ruthenia, in the Ukraine.

This is another situation where more than one "discoverer" is given credit by different references. G. W. Osann was also given credit for discovering ruthenium in platinum ore in 1827.

COMMON USES

Ruthenium's main use is as a hardener for platinum and palladium to make more serviceable jewelry and for other special purposes, including coloring glass and ceramics.

It is also used as an alloy to make electrical contacts harder and wear longer, for medical instruments, corrosion-resistant alloys, and, more recently, experimentally as solar cell material.

Ruthenium is used as a catalyst that affects the speed of chemical reactions, but is not altered by the chemical process. It is also used as a drug to treat eye diseases.

EXAMPLES OF COMMON COMPOUNDS

There are not many compounds for ruthenium. Its main use is as an alloy with other metals.

Ruthenium chloride ($RuCl_3$), also known as ruthenium trichloride—Used for technical analysis in chemistry laboratories. Highly toxic.

HAZARDS

Ruthenium powder or fine dust can be explosive when raised to high temperatures.

$RuCl_3$ is poisonous as well as highly volatile. Its fumes should be avoided.

❖ ❖ ❖

RHODIUM (Period 5)
SYMBOL: Rh   **ATOMIC NUMBER:** 45   **GROUP:** 9

COMMON VALENCE: 3 **ATOMIC WEIGHT:** 102.906 **NATURAL STATE:**
Solid **COMMON ISOTOPES:** No isotopes are known. **PROPERTIES:**
Rhodium is a silver-white, bright metallic element that is hard and durable. It
is somewhat nonreactive and insoluble in acids at room temperatures. It is
harder and has a higher melting point than platinum. Density: 12.41; melting
point: 1966°C; boiling point: 4500°C.

CHARACTERISTICS
Rhodium is one of the six platinum elements, i.e., Ru, Rh, Pd, Os, Ir, and
Pt.

Rh has the highest elec-
trical and thermal conductivity of the platinum group.

Rhodium makes an ex-
cellent electroplated surface
that is hard, wears well, and
is permanently bright.

ABUNDANCE
AND SOURCE
Rhodium is the seventy-
ninth most abundant ele-
ment and is found as alloy
mixtures with several ores,
including platinum ores. It
is recovered during the refining process. It is also recovered as a by-product of
copper extraction.

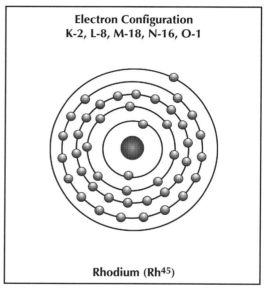

**Electron Configuration
K-2, L-8, M-18, N-16, O-1**

Rhodium (Rh⁴⁵)

These ores are found in Siberia, South Africa, and Ontario, Canada.
Rhodium is rare, but not as rare as ruthenium.

HISTORY
In the early 1800s the British chemist William Hyde Wollaston (1766–
1828) developed a method of making laboratory vessels out of platinum.
They had the desired property of being nonreactive with other chemicals.
His process was secret, and he made much money from it. At his death he
disclosed his secret.

While working with platinum, he also discovered rhodium and palladium, which he extracted from the platinum to make it more workable. He used platinum alloy mixtures of rhodium and iridium for his laboratory dishes. This provided a nonreactive metal that provided greater strength than pure platinum.

Rhodium's name is derived from the rose color of some of its compounds left over after the platinum had been removed.

COMMON USES

The most important use of rhodium is as an alloy with other metals to form high-temperature equipment, **thermocouples**, electrical contacts, and laboratory **crucibles**.

Because of its reflective properties, rhodium is used to electroplate jewelry, silverware, optical instruments, mirrors, and reflectors in lighting devices.

When combined with platinum, rhodium is used as the catalyst in the **catalytic converters** installed in automobile exhaust systems to convert toxic fumes to less harmful gases.

EXAMPLES OF COMMON COMPOUNDS

Rhodium chloride ($RhCl_3$), a red compound, insoluble in water—Used to manufacture rhodium trifluoride.

Rhodium carbonyl chloride [$Rh(Co)_2Cl_2$], reddish crystals—Used as a catalyst for organic reactions (it speeds up the reactions of organic chemicals).

HAZARDS

The powder and dust of rhodium metal are flammable in air.

Some of the compounds may cause skin irritations. It is best to use approved laboratory procedures when handling any of the six platinum family elements.

PALLADIUM (Period 5)
SYMBOL: Pd   **ATOMIC NUMBER:** 46   **GROUP:** 10

COMMON VALENCE: 2, 3, and 4 **ATOMIC WEIGHT:** 106.4 **NATURAL STATE:** Solid **COMMON ISOTOPES:** Palladium has 6 stable isotopes: palladium-102, palladium-104; palladium-105; palladium-106; palladium-108; and palladium-110. **PROPERTIES:** Palladium is a soft, white **ductile** metal that resembles platinum. It is **malleable** and can easily be rolled into thin sheets or drawn into fine wires. It does not corrode. One of palladium's unique qualities is its ability to absorb hydrogen gas to a much greater extent than other gases. Density: 12.0; melting point: 1554°C; boiling point: 2800°C.

CHARACTERISTICS

Palladium is inert to acids at room temperatures, but will react at high temperatures.

It is the most reactive of the six platinum elements: Ru, Rh, Pd, Os, Ir, and Pt.

It can absorb over 800 times its own volume of hydrogen, which gives it unique industrial properties.

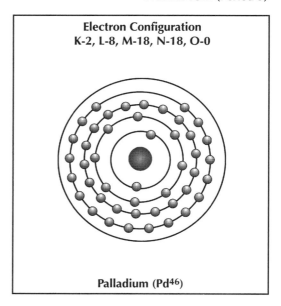

Electron Configuration
K-2, L-8, M-18, N-18, O-0

Palladium (Pd⁴⁶)

ABUNDANCE AND SOURCE

Palladium is the seventy-seventh most abundant element on Earth and is found in platinum ores along with several other platinum-like elements. It is recovered while extracting platinum from its ore. It occurs in both pure forms and in compounds of minerals.

There are deposits in Siberia and the Ural Mountains of Russia, Canada, and South Africa.

HISTORY

William Hyde Wollaston (1766–1828) discovered palladium while he was working with platinum in 1803. It can be found in both platinum and nickel ores.

Palladium was named after Pallas, an asteroid that was discovered about the same time as the element.

COMMON USES

Palladium's main uses are as alloys for special electrical switching equipment in telecommunications.

Because of its unique property of absorbing large amounts of hydrogen gas, it makes an excellent catalyst for **cracking** petroleum fractions, for **hydrogenation** of liquid oils into solid fats, and to purify hydrogen gas.

Palladium is also used to make jewelry, spark plugs, and special wires, and as a protective coating for other metals. Because it is noncorrosive, it makes excellent dental crowns.

EXAMPLES OF COMMON COMPOUNDS

Palladium chloride (PdCl$_2$)—Used to coat other metals without the need for **electrolysis**. Also used in photography, to make indelible inks, and as a catalyst in analytical chemistry (used to speed up or slow down chemical reactions).

Palladium sodium chloride (NaPdCl$_2$)—Used to test for the presence of gases, such as carbon monoxide, illuminating and cooking gas, and ethylene, and the presence of iodine.

HAZARDS

Several of the palladium compounds are oxidizing agents and some of its compounds will react violently with organic substances.

❖ ❖ ❖

SILVER (Period 5)
SYMBOL: Ag   ATOMIC NUMBER: 47   GROUP: 11

COMMON VALENCE: 1 **ATOMIC WEIGHT:** 107.868 **NATURAL STATE:** Solid **COMMON ISOTOPES:** There are 25 isotopes of silver, two of which are stable. Their atomic weights range from 102 to 117. **PROPERTIES:** Silver is a soft, white, lustrous metal that can be worked by pounding, drawing, rolling, and so forth. It is only slightly harder than gold. Silver is insoluble in water, but is soluble in several hot, concentrated acids. Density: 10.53; melting point: 961°C; boiling point: 2212°C.

CHARACTERISTICS

Chemically, silver is similar to the heavy metals. It is sometimes classed as a noble metal.

Silver is somewhat rare and is considered a commercially precious metal with many uses.

Silver has the highest electrical and thermal conductivity of all metals.

ABUNDANCE AND SOURCE

Silver is the sixty-sixth most abundant element on

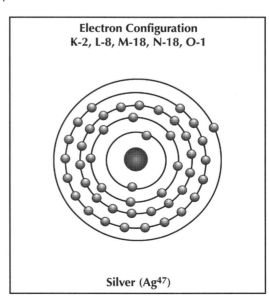

**Electron Configuration
K-2, L-8, M-18, N-18, O-1**

Silver (Ag^{47})

Earth and is sometimes found as a free metal in nature. When found pure, it is referred to as "native silver."

The major silver ores are argentite, cerfide (or cerargyrite—silver sulfide), and the sulfides of other minerals.

Silver is also derived as a by-product from the ores of other metals, including copper, lead, zinc, and gold.

HISTORY

Silver is probably one of the first metallic elements used by humans. It was known to primitive humans before 5000 B.C., at about the time copper and gold were also found in native forms. Silver jewelry was found in tombs that are over 6000 years old.

It wasn't until about 4000 B.C. that humans learned how to obtain these base metals from ores by using heat. Most likely by accident at first, and then by observation, they learned to smelt the ores to produce the metals they desired. *Argentum* is the original Latin name for silver, thus the symbol Ag.

In the mid-1800s, Thomas Wilberger Evans (1823–1897) introduced the practice of using an **amalgam** of silver, which is an alloy mixture of mercury, tin, and silver, to fill decayed teeth.

COMMON USES

Silver has many practical applications, both as the elemental metal and in many of its compounds.

It is used for dental fillings. It is also used to silver mirrors, as a catalyst for chemical reactions (to speed up or slow down the reactions), in water purification, in special batteries, and in solar cells.

Several silver compounds are important in the photography industry, in medicine, and in the jewelry industry. Sterling silver is 92.5% pure metallic silver and 7.5% copper.

EXAMPLES OF COMMON COMPOUNDS

Silver bromide (AgBr)—Used to coat photographic film and plates.

Silver chloride (AgCl)—Used in photography, to coat and silver glass, as an antiseptic, and to absorb **infrared** light in lenses.

Silver nitrate ($AgNO_3$), colorless, but turns black when exposed to light; the most important compound of silver—Used for photographic films, inks, silver plating, and as an antiseptic on wounds.

Silver nitride (Ag_3N), a colorless powder that is very sensitive to shock; will explode when disturbed, even under water—Used in percussion caps to set off explosives on contact.

Argyrol, trade name for a compound of silver and a protein—Used as an antiseptic to treat specific types of bacterial infections.

HAZARDS

Silver is not toxic as a free, elemental metal, but many of its compounds, particularly the nitrogen compounds, are toxic.

Several silver salts, in particular $AgNO_3$, are deadly when ingested, even in small amounts. When ingested, the silver compounds are slowly absorbed by the body and the skin turns bluish or black, a condition referred to as "argyria."

Several silver compounds are very explosive, e.g., silver picrate, silver nitride, silver peroxide, silver perchlorate, and silver permanganate. All are used in bullets, shells, and explosives.

❖ ❖ ❖

CADMIUM (Period 5)
SYMBOL: Cd   **ATOMIC NUMBER:** 48   **GROUP:** 12

COMMON VALENCE: 2 **ATOMIC WEIGHT:** 112.40 **NATURAL STATE:** Solid **COMMON ISOTOPES:** There are 8 stable natural isotopes and 11 unstable radioactive isotopes. **PROPERTIES:** Cadmium is a soft, **malleable** and **ductile**, lustrous blue-white metal or powder. Density: 8.642; melting point: 321°C; boiling point: 767°C. Note: Cadmium is not considered a transition element in some versions of the Periodic Table.

CHARACTERISTICS

Cadmium is located between zinc and mercury in Group 12 in the Periodic Table, and its chemical properties are similar to both.

Cadmium is malleable (it can be worked into different shapes). It **tarnishes** in moist air.

ABUNDANCE AND SOURCE

Cadmium is the sixty-fifth most abundant element, but it does not occur as a free metal in nature. It

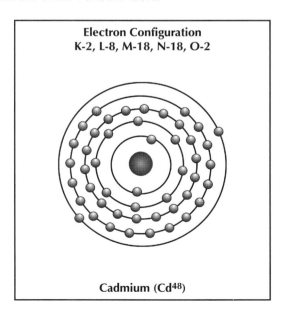

Electron Configuration
K-2, L-8, M-18, N-18, O-2

Cadmium (Cd[48])

is a rare metal. Cadmium is found only in cadmium minerals (sulfides) in nature. Almost all cadmium metal is obtained as a by-product of refining zinc, lead, and copper ores.

Greenockite (cadmium sulfide ore) also contains zinc sulfide and is found in Australia, Mexico, Peru, Zaire, Canada, Korea, Belgium-Luxembourg, and the United States.

HISTORY

In 1817 Friedrich Strohmeyer (1776–1835) analyzed some zinc ore that seemed to have another element mixed with it because, when heated, it turned yellow, which pure zinc does not do.

After separating the elements, he realized he had discovered a new element, which he named *cadmium*, Latin for "zinc ore." The name is also a derivative of the Greek word *kadmeira*, which is the word for "calamine." Calamine is commonly known as a powder or a skin lotion.

COMMON USES

Cadmium has many practical industrial uses.

Two main uses are **electroplating** and hot-dipped coatings for other metals to provide some resistance to corrosion. It is also used to make special alloys for bearings, low-temperature melting alloys, and ceramic glazes and enamels, and in welding.

It is used to make efficient nickel-cadmium storage batteries and to control the **fission** process in nuclear reactors.

Cadmium is a **neutron absorber**, so it finds use as control rods in nuclear reactors.

EXAMPLES OF COMMON COMPOUNDS

Cadmium bromide ($CdBr_2$)—Used in photography, engraving, and lithography.

Cadmium chloride ($CdCl_2$)—Used in dyeing and printing textiles, electroplating baths, photography, and as the ingredient for cadmium yellow in oil paints.

Cadmium oxide (CdO)—Used for cadmium plating baths, **electrodes** for batteries, ceramic glazes, and insecticides.

Cadmium sulfide (CdS), also called orange cadmium—Used as **pigments** in inks and paints. Also used for ceramics, **fluorescent** TV screens, **transistors**, **photovoltaic** cells, and solar cells.

HAZARDS

Cadmium powder, dust, and fumes are all flammable and toxic if inhaled or ingested. Cadmium is **carcinogenic**.

Severe illness and death can occur from exposure to cadmium and its compounds. It is absorbed from the **gastrointestinal** tract. However, it can be eliminated in the urine and feces in young, healthy people.

Cadmium, in trace amounts, is common in many of our foods; and, as we age, our bodies cannot eliminate it effectively, so cadmium poisoning may result. The symptoms of mild poisoning are burning of the eyes, irritation of the mouth and throat, and headaches. As the **intoxication** increases, there may be severe coughing, nausea, vomiting, and diarrhea. There is a 15% chance of death from cadmium poisoning.

The main risk to humans from cadmium is by industrial exposure—not from a healthy diet.

Third Series
(PERIOD 6; GROUPS 4 TO 12)

HAFNIUM (Period 6)
SYMBOL: Hf   **ATOMIC NUMBER:** 72   **GROUP:** 4

COMMON VALENCE: 2, 3, and 4   **ATOMIC WEIGHT:** 178.49   **NATURAL STATE:** Solid   **COMMON ISOTOPES:** Hafnium has 6 natural, stable isotopes. **PROPERTIES:** Hafnium is **ductile** and silvery-gray with a metallic luster. Chemically and physically, it is very similar to zirconium, which is just above it in the Periodic Table. Density: 13.1; melting point: 2200°C; boiling point varies from 2500°C to 5000°C.

CHARACTERISTICS
Hafnium is the first element of the third transition series in Period 6, which follows the lanthanide rare-Earth elements.

Hafnium has high strength and resists corrosion.

ABUNDANCE AND SOURCE
Hafnium is the forty-fifth most abundant element on Earth, and it is found in the same ores as zirconium. Zirconium is positioned just above it in Group 4 of the Periodic Table. These two similar elements are very difficult to separate and, for most of their limited uses, it is not necessary to go through the processes required to completely separate them. Hafnium is found in the minerals baddeketute (ZrO_2) and zircon ($ZrSiO_4$).

HISTORY
Early in the twentieth century several scientists were searching the "rare Earths" for new elements. According to the Periodic Table, element 72

should exist somewhere at the end of the lanthanide series and be related to other elements in Group 4. It was difficult to identify since it is always found with zirconium ore, which consists of fifty times more zirconium than hafnium.

In 1923 Georg Karl von Hevesy (1885–1966) and Dirk Coster (1889–1950), on the advice of Niels Henrik David Bohr (1885–1962), used X-ray **spectroscopic** analysis to identify element 72. They named it *hafnium,* after the Latin word "Hafnia" for Copenhagen.

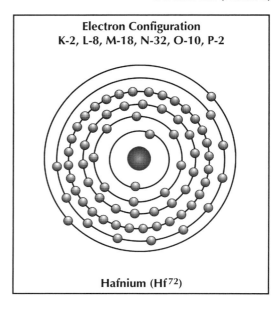

Electron Configuration
K-2, L-8, M-18, N-32, O-10, P-2

Hafnium (Hf⁷²)

COMMON USES

To make **control rods** for use in water-cooled nuclear reactors such as the type used on submarines. It is also used for **filaments** in light bulbs, as a **misch metal**, and as a **getter** to remove gases from vacuum radio tubes.

EXAMPLES OF COMMON COMPOUNDS

Hafnium carbide (HfC)—Used to make nuclear control rods.

Hafnium oxide (HfO_2)—Used for furnace refractory linings.

HAZARDS

Hafnium and its compounds are toxic if inhaled. Its dust is explosive, even when wet.

❖ ❖ ❖

TANTALUM (Period 6)
SYMBOL: Ta   **ATOMIC NUMBER:** 73   **GROUP:** 5

COMMON VALENCE: 2, 3, and 5   **ATOMIC WEIGHT:** 180.948   **NATURAL STATE:** Solid   **COMMON ISOTOPES:** Tantalum has no stable isotopes. **PROPERTIES:** Tantalum has similar properties as niobium and vanadium above it in Group 5. Tantalum is a hard, **ductile**, and **malleable** rare metal that has a bluish color when in a rough state and a silvery color if polished. Density: 16.6; melting point: 2996°C; boiling point: 5425°C.

CHARACTERISTICS

Tantalum is noncorrosive and will resist all but hydrofluoric acid.

Tantalum will ignite in air when heated.

Tantalum will form different alloys with many metals, which leads to many uses.

ABUNDANCE AND SOURCE

Tantalum is the fifty-first most abundant element on Earth. It is found free in nature, but mostly it is found in several minerals and is obtained by heating tantalum potassium fluoride or by the **electrolysis** of melted salts of tantalum.

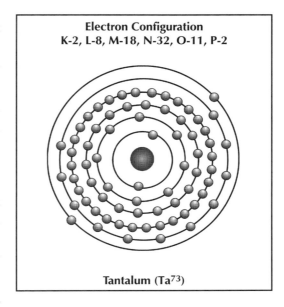

Electron Configuration
K-2, L-8, M-18, N-32, O-11, P-2

Tantalum (Ta^{73})

Ore containing tantalum (along with niobium) is found in South America, Thailand, Malaysia, and Canada.

HISTORY

Anders Gustav Ekeberg (1767–1813) discovered tantalum in 1802, while analyzing minerals sent to him from Finland by his friend Jöns Jakob Berzelius (1779–1848). Because tantalum was so difficult to separate from niobium, Ekeberg named it *tantalum* for the Greek King Tantalus, who was condemned to everlasting torment. The word means "to tantalize."

COMMON USES

The most important use is to make strong, noncorrosive alloys. Since tantalum can be drawn into thin wires, it is used in the electronics industry, to make smoke detectors, as a **getter** in vacuum tubes to absorb the air, and also as **filament** in vacuum tubes. It is also used in citizen band radios and heart pacemakers.

Because of its hardness and noncorrosiveness, it is used to make dental tools and implants, surgical equipment and surgical implants, e.g., artificial knees and hip joints; pins, screws, and plates to repair bones; and so forth.

EXAMPLES OF COMMON COMPOUNDS

Tantalum carbide (TaC)—Used to make hard and durable cutting tools.

Tantalum disulfide (TaS$_2$)—Used to make solid lubricants and special non-corrosive greases.

Tantalum fluoride (TaF$_5$)—Used as a catalyst to speed up organic chemical reactions.

Tantalum oxide (Ta$_2$O$_5$)—Used to make special optical glass, for lasers, and in electronic circuits.

HAZARDS

The dust and powder of tantalum are explosive. Several of the tantalum compounds are toxic if inhaled or ingested, but the metal is nonpoisonous.

❖   ❖   ❖

TUNGSTEN (Period 6)
SYMBOL: W   ATOMIC NUMBER: 74   GROUP: 6

COMMON VALENCE: 2, 4, 5, and 6   **ATOMIC WEIGHT:** 183.85   **NATURAL STATE:** Solid   **COMMON ISOTOPES:** There are 5 natural isotopes of tungsten. **PROPERTIES:** Tungsten is a **ductile**, but very dense and hard, brittle metal. When separated from its ores, it is a grayish solid, but it is silvery in color when purified. Tungsten has a high density: 19.3; the highest melting point of all metals: 3440°C; and a high boiling point: 5927°C.

CHARACTERISTICS

Tungsten has the highest melting point of all metals (3410°C, or 6179°F).

Chemically, tungsten is rather inert, but it will form compounds with some other elements at high temperatures, e.g., the halogens, carbon, boron, silicon, nitrogen, and oxygen. Tungsten will corrode in seawater.

ABUNDANCE AND SOURCE

Tungsten is the fifty-eighth most abundant element and is never found in

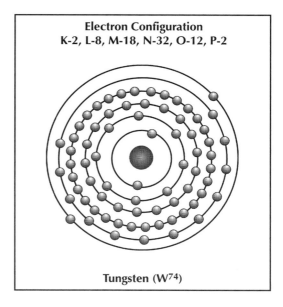

Electron Configuration
K-2, L-8, M-18, N-32, O-12, P-2

Tungsten (W^{74})

pure form in nature. Its ores (oxides) are mined in Russia, China, South America, Thailand, and Canada. In the United States it is found in Texas, New Mexico, Colorado, California, Arizona, and Nebraska.

The tungsten ore, scheelite, is named after Carl Wilhelm Scheele (1742–1786), who studied tungsten minerals, but as with many of his other "near" discoveries, e.g., oxygen, fluorine, hydrogen sulfide, hydrogen cyanide, and manganese, he was not given credit.

HISTORY

The credit for discovering tungsten as a new element is given to Don Fausto de Elhuyar (1755–1833) and his brother, who isolated a substance from tin ore they called "wolframite."

In 1783 they named it *wolfram*, from the name of the mineral. At the same time the Swedish named it *tungsten*, which means "heavy stone" in Swedish. Thus, the confused use of W for the symbol for tungsten.

COMMON USES

Tungsten has several important uses which are based on its very high melting point. It is the second most frequently used element in today's industries.

It is used to make high-strength steel that will withstand high temperatures. It makes exceptionally hard tools for cutting other materials.

Tungsten carbide is used as a substitute for diamonds for drills and grinding equipment.

It is the major metal used as the **filaments** in **incandescent** light bulbs and electronic tubes. It is used extensively in the electronics industry, for X-ray equipment, and to make solar energy equipment.

EXAMPLES OF COMMON COMPOUNDS

Tungsten carbide (WC), extremely hard and resistant to high temperatures— When cemented to tools, it is as hard as corundum (aluminum oxide) and makes excellent grinding surfaces.

Tungsten disulfide (WS_2), a solid lubricant that can withstand high temperatures—Used as a spray lubricant.

Tungsten oxide (WO_3)—Used to make tungsten alloys. Also used for fire-proofing and as a yellow pigment in ceramics.

Tungsten steel, an alloy in which tungsten acts somewhat like molybdenum to form important steel alloys; tungsten steel is tough, hard, wears well, resists rusting, and will take a sharp cutting edge.

HAZARDS

Tungsten dust, powder, and fine particles will explode, sometimes spontaneously in air.

Dust and powder forms of tungsten and tungsten compounds are toxic if inhaled or ingested.

❖   ❖   ❖

RHENIUM (Period 6)
SYMBOL: Re   **ATOMIC NUMBER:** 75   **GROUP:** 7

COMMON VALENCE: 4, 6, and 7   **ATOMIC WEIGHT:** 186.207   **NATURAL STATE:** Solid   **COMMON ISOTOPES:** There are 2 natural isotopes of rhenium. **PROPERTIES:** Rhenium ranges in color from silvery-white, to gray, to a black powder. Rhenium is a dense, **ductile** metal with a high melting point. It has one of the widest range of valences of any element. Besides the common valences of 4, 6, and 7, it also has the uncommon valences of 2, -1, and -7. Density: 21.02; rhenium has a high melting point: 3180°C, or 6220°F; and a high boiling point: 5630°C.

CHARACTERISTICS

Rhenium is one of the transition elements, which range from metals to metal-like **semiconductors**.

Its chemical properties are similar to those of technetium, which is above it in the Periodic Table, and it is not very reactive.

It forms unique **superconducting** alloys with molybdenum. It is noncorrosive in seawater.

ABUNDANCE AND SOURCE

Rhenium is the seventy-eighth most common element on Earth. It is found naturally in molybdenite and columbite ores.

The main source of rhenium is molybdenite (MoS_2), which is a natural form of molybdenum ore found in **igneous rocks** and concentrated metallic-like deposits.

Molybdenite ore is found in Chile, as well as in the states of New Mexico, Utah, and Colorado. It is also obtained as a by-product of copper and molybdenum **smelting**.

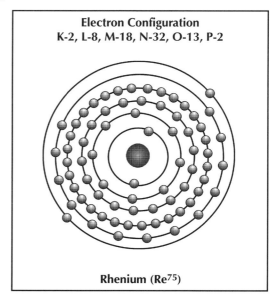

Electron Configuration
K-2, L-8, M-18, N-32, O-13, P-2

Rhenium (Re[75])

Rhenium is refined as a powder, which can then be formed by **sintering** into shapes by compressing the powder into molds under high temperatures. It is a **ductile** metal, which means it can be drawn into very thin wires and bent into various shapes.

HISTORY

Two scientists, Walter Karl Friedrich Noddack (1893–1960) and Ida Eva Tacke (Noddack) (1896–1979), discovered a new element with the atomic number of 75. They named it *rhenium*, which is the Latin word for the Rhine River.

Shortly after their joint discovery, they joined in marriage.

COMMON USES

Rhenium is used to make high-temperature alloy steels.

It is also used for electronic flash lamp **filaments** for cameras, high-temperature electrical components, electrical contacts, and **thermocouples** (two dissimilar metals joined to form a thermometer).

Rhenium is also used to make special metals for rocket and missile engines.

The radioisotope rhenium-187 is used as a standard to measure the age of the universe.

EXAMPLES OF COMMON COMPOUNDS

Rhenium heptasulfide (Re_2S_7), one of several possible compounds of rhenium and sulfur—Used as a catalyst to speed up chemical reactions and will burn when heated in air.

Rhenium heptoxide (Re_2O_7), is explosive; there are at least six different compounds of rhenium and oxygen.

Several compounds of rhenium and selenium are used in the electronics industry. Rhenium will also form compounds with some of the halogens.

HAZARDS

Rhenium is flammable in powder form. Rhenium dust, powder, and most of its compounds are toxic when inhaled or ingested.

OSMIUM (Period 6)
SYMBOL: Os   **ATOMIC NUMBER:** 76   **GROUP:** 8

COMMON VALENCE: 2, 3, 4, 6, and 8 **ATOMIC WEIGHT:** 190.2 **NATURAL STATE:** Solid **COMMON ISOTOPES:** Osmium has 7 natural, stable isotopes. They are: Os-184, Os-186, Os-187, Os-188, Os-189, Os-190, Os-192. **PROPERTIES:** Osmium is a very hard, brittle, bluish-white, rare metal in Group 8 (VIII) of Period 6 on the Periodic Table. It is not very reactive. Density: 22.59; melting point: 3045°C; boiling point: 5500°C.

CHARACTERISTICS

Osmium is one of the six transition metals in the platinum group, which includes: Ru, Rh, and Pd of the second series of transition metals (Period 5), and Os, Ir, and Pt of the third series of transition metals (Period 6).

ABUNDANCE AND SOURCE

Osmium is the eightieth most abundant element and, as a metal, it is not found free in nature; it is rare and usually found with platinum. It is recovered during the process used to extract platinum from its ore.

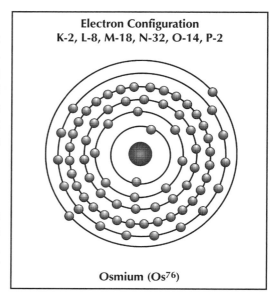

Electron Configuration
K-2, L-8, M-18, N-32, O-14, P-2

Osmium (Os^{76})

Osmium also occurs along with iridium in nature as iridosmine.

It is found in Canada, Russia, and parts of Africa.

HISTORY

Smithson Tennant (1761–1815) was working with the platinum group of metals when he discovered two new ones in 1803. One he named *osmium*, which means "smell" or "odor" in Greek; the other he named *iridium*, from the Greek word for "rainbow."

COMMON USES

When alloyed with iridium and platinum, osmium is used to make hard points for high-quality ink fountain pens and pivots for instruments such as compasses. Osmium is mixed with other metals of the platinum group to make extra-hard alloys of these metals.

Another use is as a stain for animal tissues that are to be examined with a microscope.

EXAMPLES OF COMMON COMPOUNDS

Osmium tetraoxide (OsO_4), a yellow crystal, probably the most important compound—Used as an oxidizing agent, as a stain in microscopy, and to detect fingerprints.

HAZARDS
The compound OsO_4 is extremely poisonous. Osmium is a powerful oxidizing poison that is soluble in water. It will burn holes in skin and is deadly if inhaled or ingested.

❖   ❖   ❖

IRIDIUM (Period 6)
SYMBOL: Ir   **ATOMIC NUMBER:** 77   **GROUP:** 9

COMMON VALENCE: 1, 2, 3, 4, and 6 **ATOMIC WEIGHT:** 192.22
NATURAL STATE: Solid **COMMON ISOTOPES:** There is 1 useful radioactive isotope: iridium-192, with a half-life of 74 days. **PROPERTIES:** Iridium is a hard, brittle, white, metallic substance. It oxidizes only at high temperatures and is the most corrosive-resistant element known. Also, it is highly resistant to attack by other chemicals and is one of the most dense elements found on the Earth. Density: 22.55; melting point: 2447°C; boiling point: 4500°C.

CHARACTERISTICS
Iridium is one of the platinum group of transition elements (Ru, Rh, and Pd of Period 5, and Os, Ir, and Pt of Period 6).

It is not attacked by strong acids, including aqua regia (see below). It is the only metal that can be used in equipment to withstand temperatures up to 2300°C, or 4170°F.

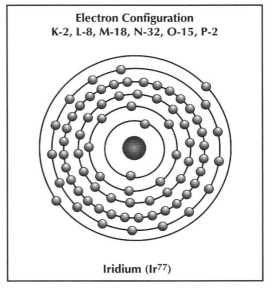

Electron Configuration
K-2, L-8, M-18, N-32, O-15, P-2

Iridium (Ir⁷⁷)

ABUNDANCE AND SOURCE
Iridium is the eighty-third most abundant element and is found alloyed with platinum. It is also a by-product of nickel and copper production.

Iridium is separated by dissolving the combined metals with **aqua regia**, which is a mixture of 1 part nitric acid and 4 parts hydrochloric acid. Aqua regia is the only acid that will dissolve platinum and gold. Once the platinum is dissolved, the iridium, which is insoluble in this strong acid, is left behind as a residue. When refined, it forms both powder and crystals.

Iridium is found in Russia, South Africa, Canada, and Alaska.

HISTORY

Smithson Tennant (1761–1815), who worked with the platinum metals, discovered both osmium and iridium. He named this new element *iridium*, for the Greek word for "rainbow," because its compounds are of many different colors. (See Osmium.)

COMMON USES

Iridium's most common use is as an **alloy** metal, which, when added to platinum, makes it harder. It is also mixed with other metals to make electrical contacts, **thermocouples** (two dissimilar metals joined to form a thermometer), and instruments that will withstand high temperatures without breaking down. It is used to make special laboratory vessels, because it won't react with most substances. An alloy of iridium and platinum is used as the standard kilogram weight because it is noncorrosive and will not change in weight over long periods of time.

The radioisotope, iridium-192, is used to treat cancer and to take X-ray pictures of metal castings to detect flaws.

Iridium is also used as a **catalyst**.

EXAMPLES OF COMMON COMPOUNDS

Iridium potassium chloride (K_2IrCl_6)—Used as a black pigment to make black porcelain kitchen and bathroom fixtures.

Iridomyrmecin ($C_{10}H_{16}O_2$), one of the few colorless compounds of iridium—Used to manufacture insecticides.

Iridosmine, not a compound, but an alloy of iridium, osmium, and a small amount of platinum—Used to make fine-pointed surgical instruments and needles, and to form fine fountain pen tips; used worldwide to make standard weights because it resists oxidation better than any other metal.

HAZARDS

The elemental metal form of iridium is almost completely inert and does not oxidize at room temperatures. But, as with several of the other metals in the platinum group, compounds of iridium are toxic. The dust and powder should not be inhaled or ingested.

❖   ❖   ❖

PLATINUM (Period 6)
SYMBOL: Pt   **ATOMIC NUMBER:** 78   **GROUP:** 10

COMMON VALENCE: 2 and 4 **ATOMIC WEIGHT:** 195.09 **NATURAL STATE:** Solid **COMMON ISOTOPES:** Platinum has 5 natural, stable isotopes. **PROPERTIES:** Platinum is a dull, silvery-white metal that is **malleable** and **ductile** and thus can be formed into many shapes. It is noncorrosive at room temperatures, and not soluble in any acid except **aqua regia** (1 part nitric acid mixed with 4 parts of hydrochloric acid). Density: 21.45; melting point: 1772°C; boiling point: 3825°C.

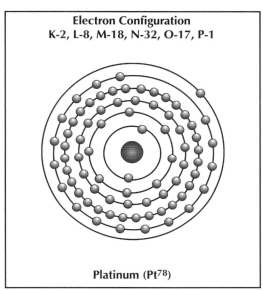

Electron Configuration
K-2, L-8, M-18, N-32, O-17, P-1

Platinum (Pt^{78})

CHARACTERISTICS

Platinum does not corrode or **tarnish**. It is the main metal of the platinum group (Ru, Rh, and Pd of Period 5, and Os, Ir, and Pt of Period 6). It can absorb great quantities of hydrogen gas and is used as a **catalyst** in industry to speed up chemical reactions.

ABUNDANCE AND SOURCE

Platinum is the seventy-fifth most abundant element and is found in pure form in nature. Like the other transition metals in the platinum group, it is widely distributed over the Earth, but it is not concentrated, which makes recovery difficult. Platinum, like gold and silver, is found as a free metal in nature, as well as in ores.

Platinum ore is treated with aqua regia, which dissolves the platinum, which is then precipitated to form platinum "sponge," which, in turn, is converted to platinum metal. Platinum is found in Russia, South Africa, South America, Ontario, Canada, and Alaska.

HISTORY

Platinum was known in ancient Egypt as early as the seventh century. A platinum metal burial-type box or casket was found in a tomb dating from that period.

Platinum was known in other countries but was not popular as a rare metal since it did not have the luster of gold or silver.

Antonio de Ulloa (1716–1795) described a free metal he found in South America that was heavier than gold, had a higher melting point, and was noncorrosive. He named it *platina*, which is the Spanish word for "silver."

An interesting bit of history. In 1978 a defector from the Communist state of Bulgaria was assassinated when he was shot by a small platinum slug that contained a deadly poison. Since platinum is inert, there was no infection or inflammation, and the physicians did not detect the wound until it was too late to save his life.

COMMON USES

The most common use of platinum is as a catalyst to speed up chemical reactions. In the early 1800s it was known that when hydrogen is passed over powdered platinum, the hydrogen ignites without being heated, and the platinum is not consumed.

Platinum is used as a catalyst in automobile emission control devices known as **catalytic converters**. The harmful gases that are the result of internal combustion engines (the gasoline engine, where the fuel burns inside the engine, as compared to the steam engine, where the fuel is burned outside the engine) are converted to less harmful gases as they pass through platinum converters. The platinum in the converter will last as long as the car since it is a catalyst and not consumed in the chemical reaction.

Platinum is also a catalyst used for **hydrogenation** of liquid vegetable oils to form solid forms of the oils. And, it is used in the **cracking** process that breaks down large crude oil molecules into smaller, more useful molecules, such as gasoline.

Platinum metal and alloys may be worked into fine wires, sheets, or special shapes. It is used to make special noncorrosive laboratory chemical reaction containers, medical instruments, jewelry, dental instruments, electric contacts, and **thermocouples**.

An alloy of platinum and iridium is used to make the standard measure for the kilogram weight. All metric weights, worldwide, are judged for accuracy against this standard kilogram.

EXAMPLES OF COMMON COMPOUNDS

Platinum black, not a compound, but a very fine powder form of platinum— Used as a catalyst and to absorb hydrogen in chemical reactions.

Platinum-cobalt, an alloy that contains about 3/4 platinum and 1/4 cobalt— Used to make the strongest permanent magnets known.

Platinum dioxide (PtO_2), a dark-brown to black powder known as Adams catalyst—Used as a hydrogenation catalyst.

Chloroplatinic acid (H_2PtCl_6), one of the most commercially important compounds of platinum—Many uses, including etching on zinc, photography, making indelible ink, plating, making mirrors, coloring in fine porcelains, and as a catalyst.

HAZARDS

Fine platinum powder or dust may explode if near an open flame. The fumes and dust of many of the compounds of platinum are toxic if inhaled or ingested. The soluble salts of platinum are very poisonous.

❖ ❖ ❖

GOLD (Period 6)
SYMBOL: Au   ATOMIC NUMBER: 79   GROUP: 11

COMMON VALENCE: 1 and 3 **ATOMIC WEIGHT:** 196.967 **NATURAL STATE:** Solid **COMMON ISOTOPES:** The only stable isotope is gold-197. The radioactive isotope, gold-198, has a half-life of 2.7 days. **PROPERTIES:** Gold is a soft, shiny, yellow, **ductile** metal (can be formed into thin wires and very thin sheets). Gold is not only pleasing to look at but pleasing to touch. Although it is chemically nonreactive, it will react with chlorine and cyanide solutions. Density: 19.3 (water = 1); melting point: 1063°C; boiling point: 2800°C.

CHARACTERISTICS

Gold is classed as a heavy, noble metal located just below copper and silver in Group 11 of the Periodic Table.

It is the most familiar of the precious metals.

About 75% of all gold is used for jewelry.

ABUNDANCE AND SOURCE

Gold is the seventy-third most abundant element and is widely spread around the world, but it is not usually found in high concentrations. It is found in pure metallic form and as compounds.

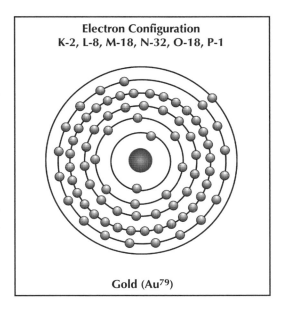

Electron Configuration
K-2, L-8, M-18, N-32, O-18, P-1

Gold (Au⁷⁹)

Seawater contains about 10 parts of gold per trillion parts (ppt) of water, which amounts to about 70 million tons in solution. In addition, there is about 10 billion tons of gold on the floors of the oceans. No economical way has been developed to recover this "treasury of the sea."

Free metallic gold is found in veins of rocks and in ores of other metals, such as quartz and pyrite. Alluvial gold (placer deposits) is found in the sand and bottom gravel of streams, where it is deposited by the movement of water over aeons.

Most gold is recovered from quartz veins called "lodes" and from ores that are crushed. The fine ore is treated with a cyanide solution that dissolves the gold. Gold is economically recovered in South Africa, Russia, Canada, Australia, Mexico, China, India, and in the states of California, Utah, Alaska, Nevada, and South Dakota. Small, scattered deposits of gold are found in several other states, including Florida, Arkansas, Washington, Oregon, Texas, Georgia, and the Carolinas.

HISTORY

Gold is considered one of the first metals used by humans. Along with other free metals, gold was known to mankind before 5000 B.C. Most likely early humans found pebble-like nuggets of metals, including gold, that they admired for their colors.

The word *metal* is from the Greek word meaning "to hunt for" or "to search for," which is what early humans did to find these rare, heavy metals. Gold was given the name *aurum*, which is Latin for "glowing dawn." Thus, the origin of the Au symbol for gold.

The search for gold was one of the contributing factors in the exploration and settlement of the New World by the Europeans in the fifteenth to eighteenth centuries.

COMMON USES

The major use of gold is for making jewelry. It finds many uses in electronics, for electrical contacts in making computer **transistors** and **diode** wires, and in dentistry. Gold leaf finds many uses in surgery, space vehicles, and art works.

Ernest Rutherford (1871–1937) used very thin gold leaf that was only 1/50,000 of an inch thick, which is only about 2,000 atoms thick, to perform his classic experiment. He bombarded the gold foil with alpha particles (helium nuclei). Most of the particles passed through the foil, but some were deflected sideways and in different directions as they were recorded on a photographic plate. This demonstrated that the atom is mostly empty space, with a small, dense, positively charged nucleus, surrounded by orbiting electrons. (See Chapter 2 for more on the structure of the atom, and Rutherford's classic experiment.)

Sodium aurothiomalate (a gold salt) is used for the internal treatment of rheumatoid arthritis, and radioactive gold-198 is used to treat cancer. Since gold is toxic when ingested, this treatment may cause complications such as skin rashes and kidney failure.

Gold is a worldwide monetary standard even though the United States went off the gold standard in the 1930s. It is still traded as a commodity. Gold coinage has a small amount of other metals added to harden the coins so they will not wear out with use. Gold bars (bullion) are stored in the treasuries of most countries. Some countries maintain huge stockpiles of gold for both monetary and industrial uses.

EXAMPLES OF COMMON COMPOUNDS

Gold cyanide [$Au(CN)_3$)], a colorless, toxic crystal that is soluble in water— Used to **electroplate** gold to metals.

Gold oxide (Au_2O_3), a dark-brown powder—Used in gold plating.

Gold alloys—Pure gold is classed as 24 karat, but it is very soft and weak. It is alloyed with other metals, such as silver, copper, nickel, and some zinc. These alloys make the gold stronger and longer wearing, but may also change the color.

The proportion of gold in jewelry is a ratio of pure gold to the alloy metals. For instance, if the jewelry is only 75% gold, it is sold as 18k (18/24). If it is sold as 14k (14/24) gold, the percentage is only about 58.33% pure gold, and 10k gold is only 41.67% pure gold.

HAZARDS

Gold, if ingested, can cause skin rash or even a sloughing off of skin. It can also cause kidney damage and possible problems with the formation of white blood cells.

MERCURY (Period 6)
SYMBOL: Hg   **ATOMIC NUMBER:** 80   **GROUP:** 12

COMMON VALENCE: 1 and 2 **ATOMIC WEIGHT:** 200.59 **NATURAL STATE:** Liquid   **COMMON ISOTOPES:** Mercury has 4 stable isotopes and 12 artificial radioactive isotopes.   **PROPERTIES:** Mercury, sometimes referred to as "quicksilver," was given the Latin name *hydrargyrum* (Hg). It is a noncombustible, heavy metal that is a liquid at room temperature. Mercury is not always included in the third series of the transition metals. Density: 13.59; freezing point: –38.85°C; boiling point: 356.6°C.

CHARACTERISTICS

Mercury is located in Group 12 (IIB) below Zn and Cd.

Mercury is not soluble in water, but will dissolve in nitric acid. It is slightly volatile.

Mercury has a high electric conductivity.

Mercury will not "wet" a surface, owing to its high surface tension, which accounts for its breakup into tiny droplets when poured over a surface.

**Electron Configuration
K-2, L-8, M-18, N-32, O-18, P-2**

Mercury (Hg⁸⁰)

ABUNDANCE AND SOURCE

Mercury is the sixty-eighth most abundant element; it does occur in its natural state but is more commonly found as a sulfide of mercury.

The most important ore of mercury is cinnabar (HgS), which has a deep-red color and is used as a pigment. It is mined in Italy, Spain, and California.

It is also found in black metacinnabar and mercury chloride. In high-grade deposits small liquid droplets of mercury may be visible. Mercury ores are found in Algeria, Mexico, Bosnia, Spain, and Canada.

HISTORY

It is most likely that mercury, like gold and silver, was known to early humans before other metals, because it could be found in a free state in nature. Many years later, copper was the first metal to be **smelted** from its ore, and many years after that, lead, tin, and iron were separated from their ores.

Mercury was used by the ancient alchemists as one of the ingredients for preparing the "philosophers' stone," which they used to try to turn base metals into gold. Alchemist/physicians used mercury to make "elixirs" to try to cure all illnesses, but often killed the patients since mercury is a poison.

Joseph Priestley (1733–1804) heated mercury with air and formed a red powder (mercuric oxide) that, when heated in a test-tube, produced small globs of mercury metal on the inside of the glass tube, as well as a gas that caused other substances to burn more rapidly than they did in air. Priestley did not know it at the time, but he had separated oxygen from the compound HgO.

Mercury was named after *Mercurius*, who, in Roman myth, was the god of travel and the messenger to other gods.

COMMON USES

One of the most common uses of mercury is to make **amalgams**, which are solutions of various metals that can be combined without melting them together. Metals such as gold, silver, platinum, uranium, copper, lead, potassium, and sodium will form amalgams with mercury. The most common is silver-mercury amalgam, used as fillings for tooth cavities.

Another common use is as a liquid contact in electrical mercury silent switches. Mercury is used for mercury-vapor and arc lamps, thermometers and barometers, and to coat mirrors.

The uses of mercury are becoming more restrictive because of its toxicity.

EXAMPLES OF COMMON COMPOUNDS

Mercuric oxide (HgO)—There are two forms, red and yellow mercuric oxides, as well as mercurous oxide (Hg_2O), which is black. All have industrial uses ranging from antiseptics to pigments.

Mercuric sulfide (HgS), a fine, very bright scarlet powder—Used as a pigment. Mercuric sulfide, like most mercury compounds, is deadly if ingested.

Mercurous chloride (Hg_2Cl_2)—Used by the **pharmaceutical** industry, to make fireworks, and as a fungicide to control maggots.

Mercury fulminate [$Hg(CNO)_2$]—Used to manufacture blasting caps and detonators. Very explosive.

HAZARDS

Mercury metal is a very toxic and accumulative poison (it is not easily eliminated by the body). When inhaled as vapors, ingested as the metal or compounds, and even absorbed when in contact with the skin, it can build up to deadly amounts. The fumes of most compounds are poisonous and must be avoided. If ingested, contact a poison center immediately.

If mercury metal is spilled, it needs to be carefully gathered so as not to spread the little globs, but rather combine them.

Some people are having the mercury amalgams used as fillings for tooth cavities removed because of the potential toxicity of the filling. There are conflicting data concerning the extent of risk resulting from the dental use of mercury amalgams, but there is some evidence that bacteria in the saliva of the mouth may leach out traces of mercury. Regardless, different metals and plastics are being used today for filling cavities.

Another danger is the waste mercury deposited by industries and agricultural chemicals in the lakes and oceans of the world. Two decades ago, most of the nations of the world approved an international ban on the dumping of mercury in our waterways and oceans. The problem is that smaller ocean plants and animals consume mercury, larger fish consume them, and then we consume the larger animals. In turn, we receive an abundance of accumulated mercury that has built up in our seafood chain. Most nations, including the United States, have banned the use of mercury agricultural chemicals, including most pesticides and insecticides.

In addition to being poisonous, some compounds of mercury are extremely explosive.

6

METALLICS AND METALLOIDS

The Periodic Table classifies elements into families that have some similar physical properties and chemical characteristics. These families of related elements are called Groups. The next three chapters will present elements from Group 13 through Group 18. The similarities and differences between the elements in each of these Groups are not uniform and require some study to be understood in relation to the Periodic Table.

Even though the elements listed in Groups 13 to 18 have a distinct number of electrons in their outer valence shells for their specific Group, they do not exhibit the same physical properties or chemical characteristics as do some of the others in the same Group. For instance, Group 13 elements have 3 electrons in their outer valence shell, while Group 14 elements have 4 electrons. Group 15 elements have 5 electrons in their outer valence shell, and Group 16 elements have 6 electrons in their outer valence shell. The halogens, in Group 17, all have 7 electrons in their outer valence shell. The inert gases, in Group 18, have a completed outer valence shell with 8 electrons, following which a new Period begins in the Table.

This does not mean that elements in each Group exhibit similar physical properties or chemical characteristics of their group. Some are more like metals and give up electrons; some are more like metalloids (semiconductors), which means they may act somewhat like metals; and some are more like nonmetals because they gain electrons.

As they progress from Group 13 to Group 17, the elements show a marked shift from metallic characteristics to properties of the nonmetals. But the distinctions are not cut-and-dried. Some elements listed in Groups 13, 14, 15, and 16 may have both metal-like elements—metalloids or semiconductors—as well as a few nonmetals.

There are several general categories of elements in Groups 13 to 17, which range from **metallics** $\rightarrow$ to $\rightarrow$ **metalloids** $\rightarrow$ to$\rightarrow$ **nonmetals**. The metallics are more metal-like than they are non-metal-like. The metalloids are now considered as semiconductors. We are using the old term "metalloids" because these elements do have characteristics of both metals and nonmetals, while the term "semiconductor" refers only to particular elements somewhere between metals and nonmetals. Semiconductors are somewhere between metals and nonmetals because they have some ability to act as conductors of electricity and thermal energy (heat), as well as the ability to act as **insulators** or nonconductors of electricity and heat.

Some examples of elements exhibiting characteristics of semiconductors or metalloids are: arsenic, antimony, silicon, germanium, polonium, phosphorus, selenium, and tellurium. Also, some organic carbon compounds can act as semiconductors.

As we move from Group 15 to 16 to 17, some of the elements at the top of each group are definitely more nonmetal in their characteristics, e.g., they gain electrons and form negative ions.

Examples of elements exhibiting "nonmetal" characteristics are the five halogens (Group 17), the five noble gases plus helium (Group 18), nitrogen, oxygen, and sulfur.

Hydrogen is an unusual element because it can be classed as either a metal or a nonmetal, depending on whether it gives up its single electron to become a positive ion or gains an electron to become a negative ion, which is called a halide when combined with another metal, e.g., lithium hydride (LiH).

These classifications (metallics, metalloids, and nonmetals) are not standard, and different references list them according to different characteristics.

Metalloids might also be thought of as "metal-like" because they exhibit both **ionic** and **covalent bonding** and will conduct electricity to some extent.

Metalloids are less **electronegative** than the nonmetals but, in general, exhibit higher electronegativity than the transition elements. They also do not give up or share electrons as readily as do the alkali metals (Group 1) and the alkali Earth metals (Group 2), i.e., they are less reactive. Metalloids are also less electronegative than are the nonmetals in Group 16 and the halogens in Group 17.

Although some metalloids do, to some extent, conduct electricity and heat, they do so less effectively than do the metals. As mentioned, metalloids will, under certain circumstances, conduct electricity. Therefore, they are often called semiconductors. Elements listed as semiconductors or metalloids are crystalline in structure. As very small amounts of impurities are added to their crystal structure, their capability of conducting electricity increases. These impurities act as electrons to carry electric currents. The flow of electricity is restricted according to the degree and types of impurities. This is why they are called "SEMIconductors."

When listed as semiconductors, not all metalloids are located in Group 13. Some are in Group 14, and a few are found in Groups 15 and 16.

In contrast, nonmetals are characterized by being very inefficient at conducting both electricity and heat. In fact, most can be thought of as "insulators" because they are such poor conductors of electricity and heat.

To confuse matters even more, some references list "semiconductors" as a special group of nonmetals.

All nonmetals except the noble gases, which have a completed outer valence shell of eight electrons, are highly electronegative. When they combine with other metals to form compounds, they either gain electrons in their outer valence shells to form ionic bonds or share electrons to form covalent bonds. When gaining electrons, the atoms become negative ions. When in **electrolytic solutions**, these ions are called *anions* because they are attracted to the positive pole (**anode**). In contrast, metals in electrolytic solutions form *cations*, which are positive ions attracted to the negative pole (**cathode**).

Again, not all of the metallics (metal-like), metalloids, and nonmetals are listed in any one Group. They can be found in Groups 13, 14, 15, 16, and 17. But the elements in each of these Groups all have the same number of electrons in their outer valence shells, e.g., Group 16 elements all have 6 electrons in their outer valence shell and Group 17 elements have 7 electrons.

We have combined the elements in Groups 13 and 14 in this chapter for convenience in discussing their metallic or metal-like characteristics, as well as their metalloid or semiconducting properties. The elements in these groups need to be individually studied and considered in the broader context of their physical and chemical nature.

Some of the elements in Group 13, **metallics,** all have 3 electrons in their outer valence shell. A few exhibit metal-like characteristics of metals. For example, aluminum can lose 1 or 3 of its valence electrons and become a positive ion just as do other metals, while other elements in this group have characteristics more like metalloids or semiconductors.

The elements in Group 14, **metalloids**, all have 4 electrons in their outer valence shell. They can form covalent bonds with other atoms and also exhibit characteristics of metalloids or semiconductors.

Elements located lower in Group 14; e.g., tin and lead, can lose their valence electrons, thus exhibiting metallic or metal-like qualities.

Elements in both Groups 13 and 14 can form ionic and covalent compounds with other elements by giving up or sharing valence electrons.

Group 13 Elements
THE METALLICS

BORON (Metallics)
SYMBOL: B   **ATOMIC NUMBER:** 5   **PERIOD:** 2

COMMON VALENCE: 3 **ATOMIC WEIGHT:** 10.811 **NATURAL STATE:** Solid   **COMMON ISOTOPES:** Two stable isotopes are known: boron-10 and boron-11.   **PROPERTIES:** Boron is dark-brown to black with a metallic luster. In the free form, it is either an **amorphous** powder or crystals. Boron is the only nonmetallic element that has less than four electrons in its outer shell. Chemically, it is more closely related to carbon (Group 14) than to aluminum (Group 13). Density: 2.45; melting point: 2100°C; boiling point: 2500°C.

CHARACTERISTICS

Boron is classified in Group 13, and is sometimes considered a nonmetal. However, since it has only 3 electrons in its outer shell (not 4 or more as do most nonmetals), it also acts more as a metalloid.

Boron is less reactive than the elements below it in Group 13. It has a high capacity to absorb neutrons. Although not all of boron's

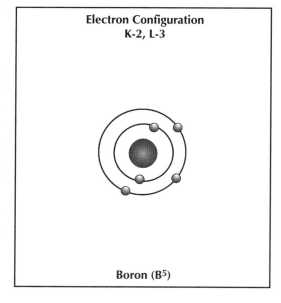

Electron Configuration
K-2, L-3

Boron (B5)

PERIODIC TABLE OF THE ELEMENTS

GROUPS PERIODS	1 IA	2 IIA	3 IIIB	4 IVB	5 VB	6 VIB	7 VIIB	8	9 VIII	10	11 IB	12 IIB	13 IIIA	14 IVA	15 VA	16 VIA	17 VIIA	18 VIIIA
1	1 H 1.0079																	2 He 4.00260
2	3 Li 6.941	4 Be 9.01218											5 B 10.81	6 C 12.011	7 N 14.0067	8 O 15.9994	9 F 18.9984	10 Ne 20.179
3	11 Na 22.9898	12 Mg 24.305											13 Al 26.9815	14 Si 28.0855	15 P 30.9738	16 S 32.066(6)	17 Cl 35.453	18 Ar 39.948
4	19 K 39.0983	20 Ca 40.08	21 Sc 44.9559	22 Ti 47.88	23 V 50.9415	24 Cr 51.996	25 Mn 54.9380	26 Fe 55.847	27 Co 58.9332	28 Ni 58.69	29 Cu 63.546	30 Zn 65.39	31 Ga 69.72	32 Ge 72.59	33 As 74.9216	34 Se 78.96	35 Br 79.904	36 Kr 83.80
5	37 Rb 85.4678	38 Sr 87.62	39 Y 88.9059	40 Zr 91.224	41 Nb 92.9064	42 Mo 95.94	43 Tc (98)	44 Ru 101.07	45 Rh 102.906	46 Pd 106.42	47 Ag 107.868	48 Cd 112.41	49 In 114.82	50 Sn 118.71	51 Sb 121.75	52 Te 127.60	53 I 126.905	54 Xe 131.29
6	55 Cs 132.905	56 Ba 137.33	★	72 Hf 178.49	73 Ta 180.948	74 W 183.85	75 Re 186.207	76 Os 190.2	77 Ir 192.22	78 Pt 195.08	79 Au 196.967	80 Hg 200.59	81 Tl 204.383	82 Pb 207.2	83 Bi 208.980	84 Po (209)	85 At (210)	86 Rn (222)
7	87 Fr (223)	88 Ra 226.025	▲	104 Und (261)	105 Unp (262)	106 Unh (263)	107 Uns (264)	108 Uno (265)	109 Une (266)	110 Uun (267)	111 Uuu (272)	112 Uub	113 Uut	114 Uuq	115 Uup	116 Uuh	117 Uus	118 Uuo

TRANSITION ELEMENTS

6 ★ Lanthanide Series (RARE EARTH)

57 La 138.906	58 Ce 140.12	59 Pr 140.908	60 Nd 144.24	61 Pm (145)	62 Sm 150.36	63 Eu 151.96	64 Gd 157.25	65 Tb 158.925	66 Dy 162.50	67 Ho 164.930	68 Er 167.26	69 Tm 168.934	70 Yb 173.04	71 Lu 174.967

7 ▲ Actinide Series (RARE EARTH)

89 Ac 227.028	90 Th 232.038	91 Pa 231.036	92 U 238.029	93 Np 237.048	94 Pu (244)	95 Am (243)	96 Cm (247)	97 Bk (247)	98 Cf (251)	99 Es (252)	100 Fm (257)	101 Md (258)	102 No (259)	103 Lr (260)

properties have been investigated, many of its compounds have practical uses in industry.

ABUNDANCE AND SOURCE

Boron is the thirty-eighth most abundant element on Earth. It makes up about 0.001% of the Earth's crust. It is not found free in nature but rather in compounds such as boron oxide.

The naturally occurring compounds are either red crystalline or less dense, dark-brown or black powder.

Boron is found in borax (a natural hydrated sodium borate found in salty lakes or in alkali soils), kernite, colemanite, and ulexite ores. It is found in many countries, including western United States.

HISTORY

Sir Humphry Davy (1778–1829) was experimenting with **electrodes** for batteries when he discovered that some of these metal-containing compounds would break down when a current was passed through water containing a solution (**electrolyte**) of the compound. The metal was deposited on one electrode and hydrogen gas was released from the water.

He used this method to isolate barium, strontium, calcium, magnesium, and boron. Another chemist of that period, Joseph Louis Gay-Lussac (1778–1850), was experimenting along the same lines and should also be given some credit for the discovery of these elements.

Alfred Stock (1876–1946) studied the **hydrides** of some of these metal-like elements. A hydride occurs when hydrogen gains (or shares) an electron rather than losing its single electron when it combines with metals or metallic-like elements. He spent years experimenting with boron hydrides (B_2H_6 and BH_3), which were used as rocket fuels powerful enough to lift rockets into space.

COMMON USES

Boron has become an important industrial element.

As a hydride (combined with hydrogen), it is an effective "booster" to rocket fuels for spacecraft.

Its main use is as an alloy for many types of metals, particularly those requiring high strength to withstand high temperatures.

It is an excellent **neutron absorber** in nuclear reactors and an oxygen absorber in producing copper and other metals.

It is also used in cosmetics (talc powder), soaps, and adhesives; as an environmentally safe insecticide; and as a water softener.

EXAMPLES OF COMMON COMPOUNDS

Boron-10, a nonradioactive isotope of boron—Used to absorb slow neutrons in nuclear reactors. It produces high-energy alpha particles (helium nuclei) in this process.

Boron carbide (B_4C), a hard, black crystal—Used as an **abrasive** powder and to strengthen composition parts in aircraft.

Boron nitride (BN), a white powder—Used to line high-temperature furnaces, for heating **crucibles**, for electrical and chemical equipment, for heat shields on spacecraft nosecones, and to make high-strength fabrics.

Refined borax ($Na_2B_4O_7$)—Used to make laundry products such as soaps and water-softening compounds. Also used for cosmetics, body powders, and the manufacture of paper and leather. Borax is an environmentally safe natural herbicide and insecticide.

Boric acid (boracic acid) (H_3BO_3)—Used for the manufacture of glass, welding, mattress batting, cotton textiles, and as a weak eyewash solution.

HAZARDS

Powdered or fine dust of boron is explosive in air and toxic if inhaled.

Several of the compounds of boron are very toxic if ingested or if they come in contact with the skin. This is particularly true of the boron compounds used for strong insecticides and herbicides.

ALUMINUM (Metallics)
SYMBOL: Al   **ATOMIC NUMBER:** 13   **PERIOD:** 3

COMMON VALENCE: 3 **ATOMIC WEIGHT:** 26.9815 **NATURAL STATE:** Solid **COMMON ISOTOPES:** No stable isotopes are found in nature. **PROPERTIES:** Pure aluminum metal is not found in nature, only in compounds with other elements. The pure metal is soft and not very strong. It is included in Group 13 and can give up 1 or all 3 of its valence electrons. Thus, it is more like a metal than a nonmetal. It is classed as a nonferrous metal. Density: 2.708; melting point: 660°C; boiling point: 2450°C.

CHARACTERISTICS

Alloys of aluminum are light, strong, and can easily be formed into many shapes, i.e., extruded, rolled, pounded, cast, and welded.

Aluminum reacts with acids and strong alkali solutions. It forms an aluminum oxide coating that prevents its continued corrosion in the atmosphere. It is a good reflector of heat.

ABUNDANCE AND SOURCE

Aluminum is not only the third most common element on Earth, it is the most abundant metal on both the Earth and moon. Even though it is not found free in nature, it is the most widely distributed element on Earth.

Almost all rocks contain some aluminum in the form of aluminum silicate minerals found in clays, feldspars, and micas.

Bauxite (aluminum hydroxides) is the major ore

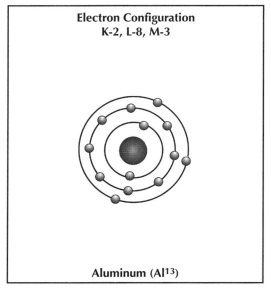

Electron Configuration
K-2, L-8, M-3

Aluminum (Al^{13})

for the source of aluminum metal. Bauxite was formed by the natural chemical reaction of water, which then formed aluminum hydroxides over the ages. A major source of bauxite, besides the United States, is Jamaica. Bauxite deposits are found in many countries, but not all are of high concentration.

HISTORY

In 1825 Hans Christian Oersted (1777–1851) was the first person to isolate a small amount of aluminum from its compounds. He made a contribution to the study of magnetism by demonstrating that when you reverse the current in a wire near a compass, the needle of the compass reverses its direction.

His process of chemical extraction by using potassium did not produce enough aluminum for practical use. Napoleon considered it a precious metal.

In 1886 Charles Martin Hall (1863–1914) was inspired by his teacher to find a way to inexpensively produce aluminum metal. He wired together numerous "wet cells" to form a "battery" to produce enough electricity to separate the aluminum from the melted ore, cryolite (aluminum oxide), by the process of **electrolysis**.

The process is now known as the "Hall-Heroult Process" because in the same year, Paul-Louis-Toussaint Heroult (1863–1914) developed the same process. Both discoverers have the same initial for their last names, the same birth and death dates, and the same date of discovery of the same process.

COMMON USES

The most common uses of aluminum are for construction, in the aviation-space industries, and in the home and automobile industries.

It is used to make cans for food and drinks, for protective coatings to resist corrosion, to manufacture die-cast auto engine blocks and parts, for home cooking foil, in **pyrotechnics**, for incendiary bombs, and for all types of alloys with other metals.

Aluminum has many **pharmaceutical** uses, including ointments, toothpaste, deodorants, shaving creams, and so forth.

Aluminum scrap is one of the salvaged and recycled metals that is less expensive to reuse than to extract the metal from its ore. It takes much less electricity to melt scrap aluminum than it does to extract aluminum from its ore.

EXAMPLES OF COMMON COMPOUNDS

Aluminum ammonium sulfate [$Al_2(SO_4)_3(NH_4)_2SO_4$]—Used as a **mordant** to fix dyes in textiles, as a chemical for water and sewage treatment, as a food additive, in pigments, and in the treatment of leather and furs.

Aluminum fluoride (AlF_3)—Used to produce low-melting aluminum metal, as a flux in ceramic glazes and enamels, and as a catalyst in **chemical reactions** (speeds up/slows down reactions).

Aluminum oxide (Al_2O_3), also the mineral bauxite, but may be found as an **abrasive**, and as rubies and sapphires—Main use is for the production of aluminum metal by electrolysis and chemical separation. Aluminum oxide is used to manufacture abrasives, ceramics, and electrical spark plug insulators. It is also used to make light bulbs, heat-resistant cloth, and artificial gems.

Aluminum alloys, although not really compounds, are by far the most important. Some other metals commonly used with aluminum are copper, manganese, silicon, magnesium, zinc, chromium, zirconium, vanadium, lead, and bismuth.

Duralumin is probably the most important alloy. It is a heat-treated mixture of aluminum, copper, magnesium, and manganese. It is very tough, strong, lightweight, and noncorrosive. It is used in the aircraft and spacecraft industries.

HAZARDS

Aluminum dust and fine powder is highly explosive and can spontaneously burst into flames in air.

Aluminum chips and course powder release hydrogen, which can burn or explode when exposed to acids. Pure aluminum foil or sheet metal can burn in air when exposed to a hot enough flame.

Fumes from aluminum welding are toxic.

❖ ❖ ❖

GALLIUM (Metallics)
SYMBOL: Ga   **ATOMIC NUMBER:** 31   **PERIOD:** 4

COMMON VALENCE: 2 and 3 **ATOMIC WEIGHT:** 69.72 **NATURAL STATE:** Solid **COMMON ISOTOPES:** There are 2 stable isotopes. **PROPERTIES:** The solid metallic gallium is bluish off-white in color and silvery in liquid form. It will liquefy at temperatures just above room temperature and become mirror-like in color. A very unique property is that it expands when solidifying from a liquid to a solid state. Gallium is chemically similar to aluminum, but has a lower melting point. It can form **monovalent** and **divalent** as well as **trivalent** compounds. Density: 5.907; melting point: 29.78°C; boiling point: 2250°C.

CHARACTERISTICS
Gallium has a wide temperature range when in the liquid state, from -35°C to 1100°C.

Gallium reacts strongly with boiling water; is slightly soluble in alkali solutions, acids, and mercury; and is used as an amalgam. It has some semiconductor properties, but has few uses in its elemental metal form.

ABUNDANCE AND SOURCE
Gallium is the thirty-fourth most abundant ele-

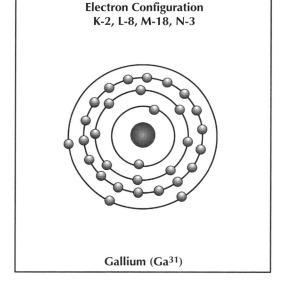

Electron Configuration
K-2, L-8, M-18, N-3

Gallium (Ga^{31})

ment but is not widely distributed as an elemental metal. It is usually combined with other elements, particularly zinc and aluminum ores.

Since 1949 the Aluminum Company of America has extracted gallium metal from aluminum. Only recently, since the development of microprocessors, chips, computers, and so forth, has gallium found many profitable uses.

HISTORY
Gallium was discovered in 1874 by Paul-Emile Lecoq de Boisbaudran (1838–1912) when he extracted gallium from zinc ore. He named it after the word *Gaul*, the Latin name for France.

Mendeleyev called it "**eka**-aluminum" because he correctly predicted gallium's properties from its placement on the Periodic Table.

COMMON USES

As a metal, gallium has very few uses, but some of its compounds (salts) are important to electronic industries.

Gallium makes a safe substitute for mercury in dental fillings when it is combined with tin and silver.

Because of its high range of temperatures as a liquid, it is used in special types of high-temperature thermometers.

Because of the unique property of some of its compounds, it is able to translate a mechanical motion into electrical impulses. This makes it invaluable for manufacturing transistors, computer chips, semiconductors, and **rectifiers**.

A unique use of gallium metal is to "glue" gemstones to metal jewelry.

EXAMPLES OF COMMON COMPOUNDS

Gallium arsenide (GaAs), **electroluminescent** in **infrared** light—Used for telephone equipment, lasers, solar cells, and other electronic devices.

Gallium phosphide (GaP), a highly pure crystal form—Used as "whiskers" and crystals in semiconductor devices.

Gallium antimonide (GaSb)—When in pure form, used in the semiconductor industries.

HAZARDS

Gallium arsenide is a highly toxic metallic compound composed of both gallium and arsenic. When it is used in the semiconductor industry, great care must be taken to protect the employees.

INDIUM (Metallics)
SYMBOL: In   **ATOMIC NUMBER:** 49   **PERIOD:** 5

COMMON VALENCE: 1 and 3 **ATOMIC WEIGHT:** 114.82 **NATURAL STATE:** Solid **COMMON ISOTOPES:** Two stable isotopes are known. **PROPERTIES:** Indium is a metal-like element that is silvery-white and softer than lead. It also can act as a semiconductor. Density: 7.31; melting point: 6156°C; boiling point: 2075°C.

CHARACTERISTICS

Indium is not found as a free element in nature.

It is chemically similar to the elements above and below it on the Periodic Table (Ga and Ti).

It is corrosion resistant at room temperature, but oxidizes at high temperatures. It is soluble in acids, but not in alkalies or hot water.

ABUNDANCE AND SOURCE

Indium is the sixty-ninth most abundant element, and it is widely spread over the Earth's crust but in very small concentrations. It is never found free in nature, and the ores of indium seldom contain more than 0.01% of the element.

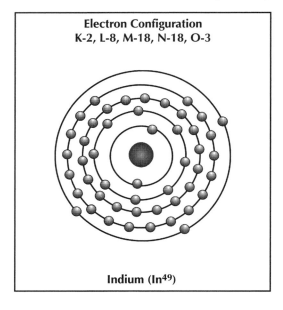

Electron Configuration
K-2, L-8, M-18, N-18, O-3

Indium (In⁴⁹)

Indium ores are found in Russia, Japan, Europe, Peru, and Canada, as well as in the western states of the United States.

Most of it is recovered from the dust and ash of industrial flues and from the residues of processing zinc.

HISTORY

Ferdinand Reich (1799–1882) found a residue in some zinc ores that he thought was a different color. Since he was color-blind, he had his assistant, Theodor Richter (1824–1898), examine it in a **spectroscope** where it showed up as a separate deep-purple line. (A spectroscope is an instrument used to analyze the wavelength of light emitted or absorbed by elements when excited. Each element has a unique line or color in the light spectrum.)

Both men received credit for the discovery and named the new element *indium* after the indigo plant found growing in India, as well as in Egypt and China.

The indigo plant was used as early as 3000 B.C. to make purple dyes. For many years humans used insects, snails, and plants to make dyes of bright red, deep red, purple, brown, yellow, and black. The processes of extracting the dyes were expensive, so only the rich and royalty could afford the deep mauve and purple dyes for their clothing and robes.

By the late 1800s many scientists tried to synthetically produce dyes such as indigo, but it was not until 1900 that synthetic indigo production was a commercial success. The synthetic production of indigo (and other synthetic dyes) destroyed the commercial agricultural trade of the extensive indigo-growing regions of India and other countries that depended on growing the plants for their livelihood.

COMMON USES

The main uses of indium are for **soldering** lead wires to semiconductors and transistors in the electronic industry.

The compounds indium arsenide, indium antimonide, and indium phosphide are used to make semiconductors that have specialized functions in the electronics industry.

Indium is used as a long-wearing coating for bearings and to produce low-melting welding and soldering alloys. It is also used as an alloy to make control rods for nuclear reactors.

Indium "wets" glass, which makes it an excellent mirror surface that lasts longer than silver or mercury mirrors.

EXAMPLES OF COMMON COMPOUNDS

Indium antimonide (InSb)—Used for semiconductors and **infrared** detecting devices.

Indium oxide (In_2O_3)—Used to manufacture special types of glass.

Indium phosphide (InP)—Used in laser devices, semiconductors, and experimentally for solar cells.

HAZARDS

Indium metal dust, particles, and vapors are toxic if inhaled. Most of the compounds of indium are also harmful if inhaled or ingested.

THALLIUM (Metallics)
SYMBOL: Tl   **ATOMIC NUMBER:** 81   **PERIOD:** 6

COMMON VALENCE: 1 and 3 **ATOMIC WEIGHT:** 204.37 **NATURAL STATE:** Solid **COMMON ISOTOPES:** Two stable isotopes are known. **PROPERTIES:** Thallium is a silvery to bluish-white metallic that is similar to lead, which is just to the right of it in Period 6, Group 14. Thallium is more like a metal than a metalloid since it forms positive ions. Density: 11.85; melting point: 302°C; boiling point: 1457°C.

CHARACTERISTICS

Elemental thallium is rare in nature.

Thallium will oxidize in air and is soluble in acids. The oxide of thallium is a black powder. Thallium is used to produce low-melting alloys and glass.

ABUNDANCE AND SOURCE

Thallium is the fifty-ninth most abundant element. It is widely distributed over the Earth, but in very low concentrations. It is found mainly in the ores of copper, iron, sulfides, and selenium.

Most of it is recovered from the flue dust of industrial smokestacks where zinc and lead ores are smelted.

HISTORY

By using **spectroscopy** in

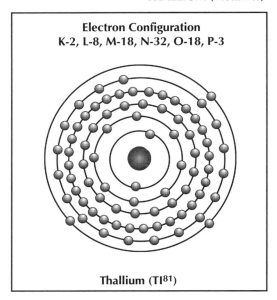

Electron Configuration
K-2, L-8, M-18, N-32, O-18, P-3

Thallium (Tl[81])

1861, William Crookes (1832–1919) was analyzing a sample of selenium ore when he saw a bright-green line in the spectrum of the spectroscope. (Spectroscopy is the analysis of elements that separates the light waves which are either given off or absorbed by elements when heated. Each element has is own unique bright line [color] or dark line [position] in the spectrum [from short to long wavelengths] of light viewed in the spectroscope.) After he analyzed this new material, he realized that he had discovered a new element, which he named *thallium*, from the Greek word *thallos*, meaning "green twig."

COMMON USES

Thallium is used as an alloy with mercury and other metals. The main uses are in **photoelectric** applications and for military infrared radiation transmitters.

It is also used to make artificial gemstones and special glass, and to make green colors in fireworks and flares. It is also used as a rat poison, as an insecticide, and to make special types of glass.

Radioactive thallium chloride (T1C1[201]) is injected into the bloodstream of heart patients during treadmill stress tests. Along with an electrocardiogram this test can detect any damaged heart tissue. Healthy heart muscle absorbs the radioactive isotope while damaged heart tissue shows up as "cold spots" on the radiation detection screen.

EXAMPLES OF COMMON COMPOUNDS

Thallium-mercury, an **amalgam**; not really a compound, but more like an alloy mixture—Used in low-temperature thermometers and as a substitute for mercury in low-temperature switches.

Thallium carbonate (Ti_2CO_3)—Used to make artificial diamonds (along with several other thallium compounds).

Thallium oxide (Ti_2O)—Used to make optical glass for microscopes, telescopes, and cameras; also to make fake diamonds.

Thallium sulfide (Ti_2S)—Used to make infrared-sensitive photoelectric cells.

HAZARDS

Thallium in all forms is very toxic if inhaled, when in contact with the skin, and, in particular, if ingested. Mild thallium poisoning causes loss of muscle coordination and burning of the skin, followed by weakness, tremor, mental aberrations, and confusion.

Thallium disease (thallotoxicosis) results from ingestion of relatively large doses (more than a few micrograms); the severity may vary with the age and health of the patient. Nerves become inflamed, hair is lost; the patient experiences stomach pain, cramps, hemorrhage, rapid heartbeat, delirium, coma, and respiratory paralysis, which will result in death in about one week. In the past thallium was one of the poisons of choice used by murderers because it acts slowly and makes victims suffer.

Group 14 Elements
THE METALLOIDS

CARBON (Metalloids)
SYMBOL: C   ATOMIC NUMBER: 6   PERIOD: 2

COMMON VALENCE: 4 **ATOMIC WEIGHT:** 12.01115 **NATURAL STATE:** Solid **COMMON ISOTOPES:** Carbon has 2 stable isotopes, carbon-12 and carbon-13, and 5 radioactive isotopes, the most important being carbon-14, which is also found in all living tissue. **PROPERTIES:** Carbon is a nonmetallic element located as the first element in Group 14. Although all elements in Group 14 have 4 electrons in their outer valence shell, carbon exhibits more nonmetal characteristics than any of the other elements in this group.

Carbon is unique in many ways. One is that it has four allotropes: (1) carbon black (jet black), (2) graphite (gray), (3) diamond (crystal clear), and (4) amorphous carbon called fullerene (no crystal structure, blackish color).

Carbon-12 is the base for determining the relative atomic weights (masses) of all the other elements. Carbon is one of only a few elements capable of forming four **covalent bonds** with other elements, giving it unique electrical properties as well as indispensability for all organic compounds.

The density and color of carbon vary, depending on the type of carbon; e.g., for graphite the density is 2.25, and for diamonds the density is 3.51. Melting point: 3500°C; boiling point: 4830°C.

CHARACTERISTICS

Carbon is, without a doubt, one of the most important elements on Earth.

It is the major element found in over one million known organic compounds.

Only a few inorganic carbon compounds exist. Elemental carbon exists in two main crystalline forms, i.e., graphite and diamonds. Carbon also exists as car-bon black, charcoal, coke, and hydrocarbons (coal, petroleum, and natural gas).

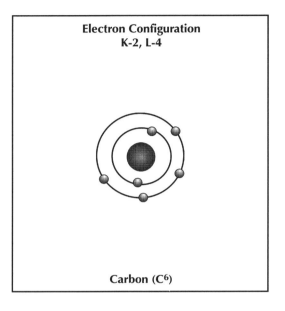

**Electron Configuration
K-2, L-4**

Carbon (C^6)

ABUNDANCE AND SOURCE

Carbon is the fifteenth most abundant element and makes up about 0.048% of the Earth's crust.

The stable isotope carbon-12 accounts for over 99% of the element carbon on the Earth and in the atmosphere. Carbon is found in all organic (living) compounds. It forms more types of compounds than all other elements combined. Many carbon compounds are combinations of carbon and hydrogen, known as hydrocarbons.

Pure carbon can be produced by burning sugar in an oxygen-free atmosphere, by burning wood to produce charcoal, or by burning coal in a low-oxygen atmosphere to produce coke.

HISTORY

Antoine-Laurent Lavoisier (1743–1794) is known as the "father of modern chemistry" because he believed in weighing, measuring, observing, heating, and testing the substances with which he experimented, and keeping accurate records of his findings. His techniques led to the field of quantitative chemistry.

Lavoisier and some other chemists pooled their funds and bought a diamond, which they placed in a closed glass jar. Using a magnifying glass, they focused the sun's rays on the diamond, which became hot enough to disappear. The weight of the glass container that held the diamond did not change, so Lavoisier then identified the gas inside as carbon dioxide, concluding that the diamond was the carbon that combined with the oxygen to form CO_2.

For years it was thought that diamonds were made of carbon atoms, just like graphite and coal, but no one could demonstrate this. In 1955 scientists were able to produce the tremendous pressure (over 100,000 times normal) and temperature (over 2500°C) to form a synthetic diamond from graphite, which is just as real as a real diamond. However, these "real" man-made diamonds are much too expensive to mass produce.

COMMON USES

There are many and varied properties and uses of carbon.

The elemental forms and compounds of carbon have more uses than all other compounds combined. **Organic chemistry**, including **photosynthesis**, is, in essence, the study of carbon chemistry. The study of **fossil fuels** is, in essence, the study of hydrocarbons.

Carbon is a strong **reducing agent**, which makes it useful in purifying metals.

Carbon, as graphite, has strong electrical conductivity properties, which makes it useful as electrodes in a variety of devices, including flashlight cells (batteries). Amorphous carbon has some superconduction capabilities.

Graphite is used for the lead in pencils, as a dry lubricant, and as electrodes in arc lamps.

Of course, carbon is very useful as jewelry, e.g., diamonds.

EXAMPLES OF COMMON COMPOUNDS

Carbon-14 is a naturally occurring radioactive form of carbon with a half-life of 5730 years. It is used to "date" any type of substance that was at one time "living." The rate of decay of carbon-14 can be calculated accurately to confirm the date when the organic substance was living.

Carbohydrates are compounds of carbon, hydrogen, and oxygen that are produced by green plants (chlorophyll) through the process of **photosynthesis**, using sunlight as an energy source. Carbon dioxide in the atmosphere is the source of the carbon. All life on Earth is dependent on carbohydrates as the ultimate source of food, and the sun is the ultimate source of all our energy, except radioactive energy.

Sucrose ($C_{12}H_{22}O_{11}$) is one of many forms of sugars (carbohydrates) that are important organic compounds for maintaining life.

Carbon dioxide (CO_2) is the eighteenth most frequently produced chemical in the United States. It has numerous uses, including in refrigeration, making carbonated drinks (soda pop), in fire extinguishers, providing an inert atmosphere, and as a **moderator** for some types of nuclear reactors.

Hydrocarbons are used as industrial fuels and sources of many chemicals. The production of the coke from coal also produces coal tars as by-products, which are used in the **pharmaceutical**, dye, food, and other industries. The refining of crude oil produces, besides gasoline, many useful petrochemicals as by-products. The range of useful products we derive from crude oil is very broad. These products not only power our automobiles, trucks, trains, and planes, but provide the base for many of our medicines, foods, and numerous other essential products.

HAZARDS

Many compounds of carbon, particularly the hydrocarbons, are not only toxic but **carcinogenic** (cancer causing).

Carbon dioxide (CO_2) in pure form will suffocate you by preventing oxygen from entering your lungs, while carbon monoxide (CO) is deadly, even in small amounts, because once you breathe it into your lungs, it takes the place of the oxygen in your bloodstream.

The elemental forms of carbon, e.g., diamonds and graphite, are not considered toxic.

❖ ❖ ❖

SILICON (Metalloids)
SYMBOL: Si   **ATOMIC NUMBER:** 14   **PERIOD:** 3

COMMON VALENCE: 4 **ATOMIC WEIGHT:** 28.086 **NATURAL STATE:** Solid   **COMMON ISOTOPES:** There are 3 natural stable isotopes: silicon-28, silicon-29, and silicon-30. All the radioactive isotopes are artificially made in nuclear reactors.   **PROPERTIES:** Elemental silicon is a dark-gray metalloid with a metallic luster. Silicon does not occur free in nature, but is found in most rocks, sand, and clay. Silicon is electropositive, so it acts like a metalloid or semiconductor. In some ways, silicon resembles metals, as well as nonmetals. In some special types of compounds called **polymers**, silicon will act in conjunction with oxygen. In these special cases it is acting like a nonmetal. Density: 2.33; melting point: 1420°C; boiling point: 3200°C.

CHARACTERISTICS

Silicon resembles the element germanium, which is just below it in Group 14 (IVA) on the Periodic Table.

Silicon in pure form is a natural semiconductor of electricity, but with some impurities it becomes an even better semiconductor.

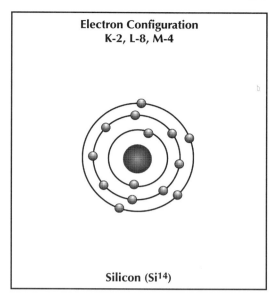

**Electron Configuration
K-2, L-8, M-4**

Silicon (Si¹⁴)

ABUNDANCE AND SOURCE

Silicon is the second most abundant element on Earth, next to oxygen. It makes up almost 28% of the Earth's crust, while oxygen makes up about 47%. Most of the rest of the crust is composed of aluminum.

Elemental silicon is produced in almost pure form by heating silicon dioxide (SiO_2), also known as silica (sand, quartz, and diatomite), along with carbon in an electric furnace.

HISTORY

In 1824 Jöns Jakob Berzelius (1779–1848), who earlier discovered cerium, ossium, and iridium, was the first person to separate the element silicon from its compounds and identify it as a new element.

Silicon research carried out in the late 1800s and early 1900s led to many forms and uses of silicon and its compounds, including silicone plastics, resins, greases, and polymers.

COMMON USES

The most important use of silicon is for **semiconductors** in the computer and electronic industry. It can be used to make solar cells to provide electricity for light-activated calculators and satellites. It has the ability to convert sunlight into electrical energy.

It is also used to form special alloys with iron, steel, copper, aluminum, and bronze. When combined with steel, it makes excellent springs for all types of uses, including automobiles.

EXAMPLES OF COMMON COMPOUNDS

The polymer (plastic) form of silicon is called silicone. It is produced in liquid, semisolid, and solid forms. Silicones are used for lubricants, coatings, water repellents for clothes, adhesives, cosmetics, polishes, sealants, electrical insulation, and surgical/reconstructive implants. Silicone polymers are used to produce many useful products.

Silicon nitride (So_3N_4), resistant to **oxidation**—Used as a coating for other metals. Also used as an adhesive, an abrasive, and to make high-temperature crucibles. Also used to make heat-resistant nozzles for rocket engines.

Silicon carbide (SiC), an abrasive—Used for grinding wheels and cutting tools. Also used to line furnaces and as a **refractory** in producing noniron metals.

Sodium silicate, also known as "water glass"—Used in the production of plastics, antacids, coatings, and cements.

HAZARDS
The dust of silicon oxide (silicate) can burn or explode and is very harmful if inhaled. Continued exposure to silica dust causes silicosis, a form of pneumonia.

The hydrides of silicon (silicon plus hydrogen) are extremely volatile and spontaneously burst into flames in air. They must be used in a special vacuum chamber.

There has been some concern over the potential hazards and safety of the cosmetic use of silicone breast implants. Clinical studies are being conducted to determine if such implants are safe. Several manufacturers have been sued by those whose implants have led to problems, which may or may not have been caused by the silicone.

❖   ❖   ❖

GERMANIUM (Metalloids)
SYMBOL: Ge   **ATOMIC NUMBER:** 32   **PERIOD:** 4

COMMON VALENCE: 2 and 4 **ATOMIC WEIGHT:** 72.59 **NATURAL STATE:** Solid **COMMON ISOTOPES:** None are known. **PROPERTIES:** Germanium is a silvery-white, brittle element that looks like a metal but is more like a nonmetal in its pure form. It is sometimes classed as a metallic, a metalloid, a nonmetal, or even, at times, as a transition element due to its unique properties. Density: 5.323; melting point: 937.4°C; boiling point: 2830°C.

CHARACTERISTICS
Germanium acts like a nonmetal in its pure form, since it does not conduct electricity. However, as impurities are added it becomes more like a metal because it then conducts electricity. This accounts for its properties as a semiconductor. Germanium is located between silicon and tin in the Peri-

odic Table and has properties of both. It is most like
silicon in its semicon-
ducting properties.

ABUNDANCE AND SOURCE

Germanium is the fifty-sec-
ond most abundant element and
is widely distributed on the
Earth. About 7 ppm of the element are found in the crust of
the Earth.

Germanium is not found
free in the elemental state in
nature but is found in the sulfides of other ores, such as
zinc, copper, lead, tin, and an-
timony. It can be found in some types of coal and ash.

Germanium has been recovered in Oklahoma, Kansas, Missouri, and
Texas.

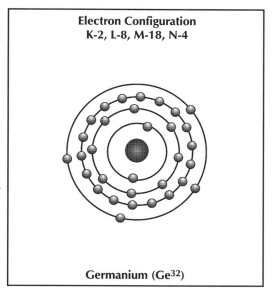

**Electron Configuration
K-2, L-8, M-18, N-4**

Germanium (Ge^{32})

HISTORY

Clemens Alexander Winkler (1838–1904) was analyzing silver ore when
he came up short on the resulting products by 7%. He kept working on this
problem, and in 1886 he found the missing 7% as a newly identified element.

Mendeleyev had accurately predicted this new element's placement in
his Periodic Table. He called it "**eka**-silicon," which was finally named
germanium after Winkler's native country, Germany.

In 1948 William Bradford Shockley (1910–), who is considered the
inventor of the **transistor**, and his associates at Bell research laboratories,
Walter Houser Brattain and John Bardeen, discovered that a crystal of
germanium could act as a semiconductor of electricity. This unique property
of germanium indicated to them that it could be used as both a **rectifier** and
an amplifier to replace the old glass radio vacuum tubes.

Their friend John Robinson Pierce (1910–) gave this new solid-state
device the name *transistor*, since the device had to overcome some resistance when a current of electricity passed through it.

COMMON USES

The most common and important use of germanium is for solid-state
devices in the electronics industry.

As a semiconductor, germanium can be used to make transistors, **diodes**, and numerous types of computer chips. It was the first element that could be designed to act as different types of semiconductors for a variety of applications. By adding small amounts of impurities to germanium, the degree of its electrical conductivity can be controlled.

Germanium is also used as a **brazing** alloy, to produce **infrared** transmitting glass and other types of lenses, and to produce synthetic garnets (semiprecious gemstones) that have special magnetic properties.

EXAMPLES OF COMMON COMPOUNDS

Germanium dioxide (GeO_2)—As 99.999% pure, is used as a **semiconductor**, in transistors, in diodes, and to make special **infrared** transmitting glass.

Germanium tetrahydride (GeH_4)—Used to produce crystals of germanium. It is extremely toxic.

HAZARDS

Germanium-halogen compounds are all very toxic, both as powder and as vapors.

Precautions should be taken when working with germanium as with similar metalloids from Group 14 (IVA).

TIN (Metalloids)
SYMBOL: Sn   **ATOMIC NUMBER:** 50   **PERIOD:** 5

COMMON VALENCE: 2 and 4 **ATOMIC WEIGHT:** 118.69 **NATURAL STATE:** Solid **COMMON ISOTOPES:** There are 10 isotopes of tin. **PROPERTIES:** Tin is a silver-white, rather soft, **ductile**, **malleable** metal that can be "pulled" into wires, hammered into shapes, and rolled into sheets. There are two major forms of stannous (tin), based on the valence—(Sn^{2+}) and (Sn^{4+})—plus many inorganic and organic compounds, i.e., with carbon. Density: 7.3; melting point: 232°C; boiling point: 2260°C.

CHARACTERISTICS

Tin is located between germanium and lead on the Periodic Table. It has characteristics similar to both, but is considered a metal because it forms positive ions in chemical reactions.

Tin has a relatively low melting point. It reacts to some acids and strong alkali, but not with hot water.

The metal tin is very workable. Because tin resists corrosion, it has many industrial uses.

ABUNDANCE AND SOURCE

Tin is the forty-ninth most abundant element. There are no high-grade tin ore deposits on Earth. The most important ore is cassiterite (SnO_2), also known as tinstone. Tin is also recovered from low-grade gravel deposits. An important source is the recovery of tin from other tin products, such as tin plate.

One of the most important sources for tin is Malaysia where it is found in profitable deposits. It is also mined in Bolivia and Thailand. Small deposits are found on the southeast coast of England.

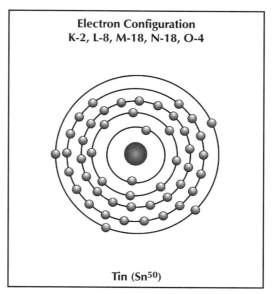

Electron Configuration
K-2, L-8, M-18, N-18, O-4

Tin (Sn^{50})

HISTORY

Tin was known and used as long ago as 3600 B.C., when it was added to melted copper to make a hard metal alloy called *bronze*. Bronze was a very important metal because of its hardness and ability to maintain a cutting edge, thus making it suitable for weapons and tools. Before tin was used to make bronze, early humans used arsenic, but this caused death to those who worked with it.

An interesting bit of history. About 500 B.C. the tin mines of the eastern Mediterranean were depleted. This created a problem since tin, being a rather rare metal, was very necessary to make the hardened bronze alloys for weapons and tools. This also is thought to be the first time in man's history that humans ran out of a mineral due to mining.

As the story goes, the Phoenicians set sail westward through the Straits of Gibraltar and discovered what were known as the "tin islands" somewhere in the Atlantic Ocean. Since this was a rich find of an essential metal, they kept the source a secret.

It is now believed that they actually found tin on the coast of England in a section we now know as Cornwall, which still produces tin ore.

The Latin name for tin is *stannum*, hence Sn for the chemical symbol.

COMMON USES

One of the most important uses of tin is to coat thin steel sheets to make "tin plate," which is then used to make what is known as the "tin can." The tin coating on the inside of the steel is thin, cheap to apply, and is resistant to most foods for long periods of time.

Tin is added to lead to make low-melting alloys for fire prevention sprinkler systems and easy-melting solder.

It is used for bearings, to plate **electrodes**, and to make pewter, **Babbitt metal**, and dental **amalgams**.

It has been used, along with other metals, for making castings for letter type in printing presses.

Some compounds of tin are used as fungicides and insecticides.

Tin is used for weighting silk, to give the fabric more body and weight.

EXAMPLES OF COMMON COMPOUNDS

Stannous oxide (SnO)—A bluish-black crystal used for metal plating and glass manufacturing.

Stannic oxide (SnO_2)—A whitish-powder used as a ceramic glaze and polishing agent.

Stannous chloride (similar to the stannous oxides)—There are two forms, stannous^{2+} chloride ($SnCl_2$) and stanic^{4+} chloride ($SnCl_4$), which are used as **electrolytes** in electrotinning processes, and as stabilizers in perfumes and soaps.

Stannous fluoride (SnF_2)—Used as a toothpaste additive to help prevent tooth decay.

HAZARDS

Tin, as the elemental metal, is nontoxic. Most, but not all, of tin's inorganic salts and compounds are also nontoxic.

In contrast, almost all organic tin compounds (those composed of carbon and hydrocarbons) are very toxic and should be avoided; if they are used, special equipment and care must be taken in handling them.

Note: When chemical formulas use the letter "R" preceding an element's symbol, it designates some form of organic compound, e.g., R_4Sn. If the letter "X" follows the element's symbol in a formula, it designates some form of inorganic compound, e.g., SnX_2. Thus, a whole series of tin compounds could be designated as R_4Sn_2, R_2Sn or SnX_4, SnX_2, and so forth.

❖ ❖ ❖

LEAD (Metalloids)
SYMBOL: Pb **ATOMIC NUMBER:** 82 **PERIOD:** 6

COMMON VALENCE: 2 and 4 **ATOMIC WEIGHT:** 207.2 **NATURAL STATE:** Solid **COMMON ISOTOPES:** Lead has 4 stable isotopes. They are the end result of millions of years of radioactive decay of three naturally radioactive elements: uranium-206, thorium-208, and actinium-207. **PROPERTIES:** Lead is a bluish-white, heavy metallic element found in Group 14 (IVA) of the Periodic Table. Its properties are more metal-like than metalloid or nonmetal. It is only slightly soluble in water and most acids, except nitric acid, with which it forms salt compounds. It is noncombustible and resists corrosion. Density: 11.35; melting point: 327°C; boiling point: 1755°C.

CHARACTERISTICS

Lead forms important industrial alloys with other metals, such as: copper, tin, arsenic, antimony, cadmium, bismuth, and sodium.

Lead can be formed into sheets, pipes, buckshot, wires, and powder. Lead is a poor conductor of electricity but an excellent shield for protection from radiation, including X-rays and gamma rays.

ABUNDANCE AND SOURCE

Lead is the thirty-sixth most abundant element on

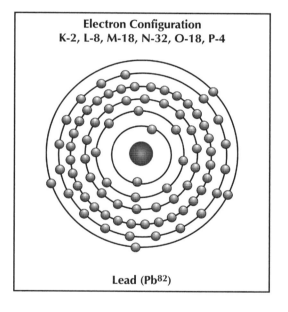

Electron Configuration
K-2, L-8, M-18, N-32, O-18, P-4

Lead (Pb⁸²)

Earth. It is rarely found in the free elemental metal state in nature. It is most often found in its sulfide state in the mineral galena. Lead is obtained by roasting (heating in oxygen to convert sulfide ores to oxides) and then reducing the oxides to metallic lead. Much lead is also recovered by recycling lead products such as automobile storage batteries.

There are several types of lead ores: lead sulfide (PbS—galena ore), lead sulfate ($PbSO_4$—anglesite ore), and lead carbonate ($PbCO_3$—cerussite ore). Lead is also found in the ore jamesonite. Lead ores are located in Europe (Germany, Rumania, and France), Africa, Australia, Mexico, Peru, Bolivia, Canada, and the United States. The largest producer of lead in the United States is the Yellow Pine mine at Stibnite, Idaho. Lead is also mined in Missouri, Kansas, Oklahoma, and Montana.

One of the most famous mining towns was the high-altitude western city of Leadville, Colorado. The boom started with the gold rush of the 1860s, followed by silver mining in the 1870s and 1880s. Today this city is the site of mining operations not only for lead, but also for zinc and molybdenum. At the height of its fame, Leadville had a population of almost 50,000 people. Today its population is about 2,500.

HISTORY

The early alchemists knew about lead, but often confused it with several other metals and their ores. The metallic forms of lead, mercury, arsenic, antimony, bismuth, and zinc were not known as separate elements until methods were developed to analyze these ores and their metals. Once the metals could be separated, they were identified as distinct elements.

The symbol Pb stands for *plumbum,* which is the Latin word for lead. This is also the derivation of the word "plumber," because plumbers work with lead pipes.

Lead-lined pipes were used by the ancient Romans to bring water from their famous aqueducts to their homes. In addition, most of the population of Rome used pots made of lead and lead alloys in which they cooked their food. Since lead is slightly soluble in water, it is possible that much of the population was poisoned, to some degree, by lead. Although there is scant evidence that mass lead poisoning existed, it has been speculated that lead pipes and cooking utensils may have helped accelerate the decline of the Roman Empire.

COMMON USES

Lead has many important commercial uses.

In the past, tetraethyl lead was added to gasoline to slow its burning rate to prevent engine knock. This caused serious and harmful air pollution, and lead has since been eliminated as a gasoline additive in most countries. Most exterior (and some interior) house paints contained high levels of lead. Today the amount of lead in paint is controlled from 0.0% to 0.05%.

Storage batteries are one of the most important uses of lead today. New combinations of metals and **electrolytes** are being explored that will increase the efficiency of storage batteries, thus eliminating the need to use lead.

Lead is used for radiation shielding, solder, bearings, cable covering, ammunition (bullets), pipes, foil, and many alloys.

EXAMPLES OF COMMON COMPOUNDS

White lead [$2PbCO_3 \cdot Pb(OH)_2$]—Used as a lead pigment in paints and other substances.

Lead arsenate [$Pb_3(AsO_4)_2$]—Used as an insecticide and herbicide. It is extremely toxic.

Babbitt metal—A soft alloy of lead with tin, cadmium, and arsenic, used as a metal for oil-less bearings.

Lead azide [$Pb(N_3)_2$]—Used as a detonator for explosives. It is extremely sensitive and must be handled with care.

Lead oxide (Pb_3O_4)—A red compound used in storage batteries and to manufacture glass, glazes, enamels, etc. Also used as a packing for pipe joints.

There are hundreds more important compounds of lead.

HAZARDS

Lead is probably one of the most widely distributed poisons in the world. Not only is the metal poisonous, but most lead compounds are also extremely toxic when inhaled or ingested. A few, such as lead alkalies, are toxic when absorbed through skin contact.

Workers in industries using lead are subjected to testing of their blood and urine to determine the levels of lead in their systems. Great effort is made to keep the workers safe.

Unfortunately, many older homes (built prior to 1950) have several coats of lead paint that flake off, which then may be ingested by children, causing various degrees of lead poisoning, including mental retardation or even death.

Young children are more susceptible to an accumulation of lead in their systems than are adults because of their smaller body size and more rapidly growing organs, such as the kidneys, nervous system, and blood-forming organs. Symptoms may include headaches, dizziness, insomnia, and stupor, leading to coma and eventually death.

Lead poisoning can also occur from drinking tap water from pipes that have been soldered with a lead alloy solder. This risk can be reduced by running the tap water until it is cold, which assures a fresher supply of water.

Another hazardous source is pottery that is coated with a lead glaze. Acidic and hot liquids (citrus fruits, tea, and coffee) react with the lead, and each use adds a small amount of ingested lead that can be accumulative. Lead air pollution is still a problem, but not as great as before, since tetraethyl lead is no longer used in gasoline. However, lead air pollution is a problem for those living near lead **smelting** operations or in countries where leaded gasoline is still used.

Even though lead and many of its compounds are toxic and **carcinogenic**, our lives would be much less satisfying without it.

7

METALLOIDS AND NONMETALS

Group 15 Elements
THE SEMICONDUCTORS

The elements in Groups 15 (VA) range from metal-like metalloids to nonmetals. Metalloids are also known as **semiconductor** elements that exhibit some characteristics of metals as well as nonmetals. A semiconductor is an element that can, to some extent, conduct electricity and heat, which is a property of metals. The ability of semiconductors to transmit electrical currents can be enhanced by controlling the number and amount of impurities. This is what makes them act as "on-off" circuits to control electrical impulses. This property is very valuable in the electronics industry for the production of **transistors**, computer chips, integrated circuits, etc. In other words, how well a semiconductor conducts electricity is not entirely dependent on the pure element itself, but also depends on the degree of its impurities and how they are controlled.

All the elements in Group 15 have 5 electrons in their outer valence shells. They can lose, gain, or share any number of these. For example, the lightest element in Group 15 is nitrogen, which, as a **diatomic** molecular gas (N_2), is not very reactive, yet it forms many compounds as an element. As an element, it exhibits valences of 1, 2, 3, 4, and 5. It can either gain electrons, share its electrons, or give up electrons with many other elements,

PERIODIC TABLE OF THE ELEMENTS

GROUPS / PERIODS

TRANSITION ELEMENTS

1 IA	2 IIA	3 IIIB	4 IVB	5 VB	6 VIB	7 VIIB	8 VIII	9 VIII	10 VIII	11 IB	12 IIB	13 IIIA	14 IVA	15 VA	16 VIA	17 VIIA	18 VIIIA
1 **H** 1.0079																	2 **He** 4.00260
3 **Li** 6.941	4 **Be** 9.01218											5 **B** 10.81	6 **C** 12.011	7 **N** 14.0067	8 **O** 15.9994	9 **F** 18.9984	10 **Ne** 20.179
11 **Na** 22.9898	12 **Mg** 24.305											13 **Al** 26.9815	14 **Si** 28.0855	15 **P** 30.9738	16 **S** 32.066(6)	17 **Cl** 35.453	18 **Ar** 39.948
19 **K** 39.0983	20 **Ca** 40.08	21 **Sc** 44.9559	22 **Ti** 47.88	23 **V** 50.9415	24 **Cr** 51.996	25 **Mn** 54.9380	26 **Fe** 55.847	27 **Co** 58.9332	28 **Ni** 58.69	29 **Cu** 63.546	30 **Zn** 65.39	31 **Ga** 69.72	32 **Ge** 72.59	33 **As** 74.9216	34 **Se** 78.96	35 **Br** 79.904	36 **Kr** 83.80
37 **Rb** 85.4678	38 **Sr** 87.62	39 **Y** 88.9059	40 **Zr** 91.224	41 **Nb** 92.9064	42 **Mo** 95.94	43 **Tc** (98)	44 **Ru** 101.07	45 **Rh** 102.906	46 **Pd** 106.42	47 **Ag** 107.868	48 **Cd** 112.41	49 **In** 114.82	50 **Sn** 118.71	51 **Sb** 121.75	52 **Te** 127.60	53 **I** 126.905	54 **Xe** 131.29
55 **Cs** 132.905	56 **Ba** 137.33	★	72 **Hf** 178.49	73 **Ta** 180.948	74 **W** 183.85	75 **Re** 186.207	76 **Os** 190.2	77 **Ir** 192.22	78 **Pt** 195.08	79 **Au** 196.967	80 **Hg** 200.59	81 **Tl** 204.383	82 **Pb** 207.2	83 **Bi** 208.980	84 **Po** (209)	85 **At** (210)	86 **Rn** (222)
87 **Fr** (223)	88 **Ra** 226.025	▲	104 **Und** (261)	105 **Unp** (262)	106 **Unh** (263)	107 **Uns** (264)	108 **Uno** (265)	109 **Une** (266)	110 **Uun** (267)	111 **Uuu** (272)	112 **Uub**	113 **Uut**	114 **Uuq**	115 **Uup**	116 **Uuh**	117 **Uus**	118 **Uuo**

★ Lanthanide Series (RARE EARTH) — Period 6

57 **La** 138.906	58 **Ce** 140.12	59 **Pr** 140.908	60 **Nd** 144.24	61 **Pm** (145)	62 **Sm** 150.36	63 **Eu** 151.96	64 **Gd** 157.25	65 **Tb** 158.925	66 **Dy** 162.50	67 **Ho** 164.930	68 **Er** 167.26	69 **Tm** 168.934	70 **Yb** 173.04	71 **Lu** 174.967

▲ Actinide Series (RARE EARTH) — Period 7

89 **Ac** 227.028	90 **Th** 232.038	91 **Pa** 231.036	92 **U** 238.029	93 **Np** 237.048	94 **Pu** (244)	95 **Am** (243)	96 **Cm** (247)	97 **Bk** (247)	98 **Cf** (251)	99 **Es** (252)	100 **Fm** (257)	101 **Md** (258)	102 **No** (259)	103 **Lr** (260)

both metals and nonmetals. The heavier elements located in the lower section of Group 15 are more likely to give up or share electrons somewhat as do metals, as compared to the lighter ones at the top of the Group that can receive electrons. The elements found in Group 15 are N^7, P^{15}, As^{33}, Sb^{51}, and Bi^{83}.

NITROGEN (Metalloids)
SYMBOL: N   **ATOMIC NUMBER:** 7   PERIOD: 2

COMMON VALENCE: 1, 2, 3, 4, and 5 **ATOMIC WEIGHT:** 14.0067
NATURAL STATE: Gas **COMMON ISOTOPES:** Nitrogen has 2 stable isotopes: nitrogen-14 and nitrogen-15: In addition, there are 4 radioactive isotopes: nitrogen-12, nitrogen-13, nitrogen-16, and nitrogen-17, which have been man-made in **nuclear reactors**. **PROPERTIES:** Although nitrogen is located in Group 15, it is considered a nonmetal. Nitrogen, as a diatomic molecule (N_2), is a colorless, odorless, tasteless gas. Gas is the natural state of nitrogen, which makes up about 78% of all the gases in the atmosphere. Density: 0.96737 (air = 1.0); freezing point: $-210°C$; boiling point: $-195.5°C$.

CHARACTERISTICS
Nitrogen is slightly soluble in water and alcohol. It is noncombustible and is considered an **asphyxiant** gas; i.e., if you breathe pure nitrogen gas it will deprive your body of oxygen.

Nitrogen is nonreactive in both the gas and liquid states. It can be liquefied at very low temperatures. At high temperatures, nitrogen reacts with many metals to form nitrides.

ABUNDANCE AND SOURCE
Nitrogen is the thirtieth most abundant element and it makes up about 2/5 of the air in our atmosphere, or 78% by volume. The balance of nitrogen with other

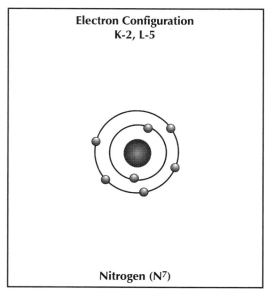

Electron Configuration
K-2, L-5

Nitrogen (N^7)

gases in the atmosphere is maintained by the nitrogen cycle, which includes nitrogen fixation of bacteria in the soil by **legumes** (bean and pea plants).

Lightning produces nitrogen, as do industrial waste gases and the decomposition products of organic material, i.e., organic proteins and amino acids in plants and animals, contain nitrogen. These sources replace the nitrogen in the atmosphere to complete the cycle.

Ammonia (NH_3) is the first binary molecule discovered in outer space of our galaxy, the Milky Way. It may also be the main compound that forms the rings of the planet Saturn.

HISTORY

In 1772, a student, Daniel Rutherford (1749–1819), at the suggestion of his mentor, Joseph Black (1728–1799), conducted an experiment where he burned a candle in a closed container of air. (Joseph Black was famous for his concept of "fixed air," which was an important step in understanding gases in chemistry.)

Rutherford noticed that there was still a large quantity of gas in the container after the burning candle had consumed the oxygen. It was not carbon dioxide since he had chemically removed that gas. Rutherford experimented further and found that this leftover gas could not support combustion or support life, so he called it *noxious air*. He did not know that he had discovered a new element, which was later named *nitrogen* for the Greek word meaning "niter producer" after the compound, potassium nitrate, which contains nitrogen.

At the same time as Rutherford was conducting his experiments, three other chemists—Priestley, Cavendish, and Scheele—were also investigating "fixed air" gases, including nitrogen. However, Rutherford was given credit for discovering nitrogen.

COMMON USES

Nitrogen has many uses. It is the second most commonly produced chemical in the United States. Some uses are based on the inertness of the gas (N_2) and its ability to be liquefied to provide very low temperatures. Other uses are based on its multiple valences, which, when nitrogen is combined with several other elements, makes it useful as fertilizers and explosives.

As a gas from the atmosphere, it is used to produce anhydrous ammonia (NH_3), which is the fifth most commonly produced chemical in the United States. It is used to make many nitrogen compounds.

Nitrogen turns into a liquid at -210°C, which makes it excellent for fast freeze-drying foods and as a **cryogenic** supercoolant.

Probably the most important use of nitrogen occurs in nature, in *nitrogen fixation*, where atmospheric nitrogen is "fixed" by bacteria to form nitrates which are absorbed by the root hairs of plants. This nitrogen enables plants (mostly legumes) to **synthesize** proteins, which become a major part of our diet. The nitrogen cycle is completed when plants and animals decay and replace nitrogen in the air they obtained from food.

Breathing compressed "air," which is a mixture of nitrogen and oxygen, can cause some serious physical symptoms when deep-sea diving or scuba diving. If divers go too deep for long periods of time, the nitrogen in the compressed-air tanks can cause "nitrogen narcosis," which causes the divers to become disoriented. And, if the divers come to the surface too fast, nitrogen gas bubbles may form in the blood faster than the bubbles can be absorbed into the body, causing what is known as "the bends." This is why a mixture of helium and oxygen is used in air tanks. The helium is more soluble than nitrogen in the blood, thus reducing the chance for the bends to occur.

Nitrogen is essential to living things. Plants require nitrogen and so do animals, which get their nitrogen from eating plants or other animals.

EXAMPLES OF COMMON COMPOUNDS

Ammonia (NH_3) is a very important compound of nitrogen. Either as a gas or as ammonium nitrate, it is used as a fertilizer and as a component for explosives. Ammonia is also used to produce several other compounds, such as nitric acid, hydrogen cyanide, urethane, and sodium carbonate.

Ammonium hydroxide (NH_4OH) is a colorless liquid that, with its strong odor, is irritating to the eyes. We all know weak solutions of it as household cleaning ammonia. Concentrated ammonium hydroxide has many industrial uses, including the manufacture of rayon, fertilizers, refrigeration, rubber, **pharmaceuticals**, soaps, **lubricants**, ink, explosives, and household cleaners.

Ammonium nitrate (NH_4NO_3), also known as "Norway saltpeter," is mainly used as a fertilizer. It is also known as the chemical that was mixed with diesel fuel to create the explosion that demolished the Federal Building in Oklahoma City in 1995.

Nitrogen oxides (NO_x) (the "x" represents the proportion of nitrogen to oxygen atoms as 1 to 5, based on the multiple valences) have many uses, including the production of nitric acid. Most of the oxides of nitrogen, especially NO_2, are toxic if inhaled. On the other hand, nitrous oxide (N_2O), although explosive in air, is known as "laughing gas" and is used as an **anesthetic** in dentistry and surgery.

Nitroglycerin is a complicated formulation of nitrogen that is very shock sensitive and is used as a high explosive. When mixed with inert matter, it can be stabilized as dynamite, which requires a blasting cap to detonate it. Nitroglycerin is also used as a **vasodilator** to reduce high blood pressure and angina pectoris by dilating the blood vessels of heart patients whose hearts are not receiving an adequate blood supply.

HAZARDS
Nitrogen is nontoxic, but it is an asphyxiant gas that cannot, by itself, support **oxidation** (combustion) or support life. If you breathe pure nitrogen for any period of time, you will die—not because the nitrogen gas is a poison (it is not) but because your body will be deprived of oxygen.

Many of the nitrogen compounds used to make fertilizers are also used in making explosives, some of which can be extremely dangerous because they are very sensitive to shock or heat.

Several of the oxygen, hydrogen, and halogen compounds of nitrogen are toxic when inhaled. A common error made in using household cleaners is to mix or use together ammonia cleaning fluids (containing nitrogen) and Clorox-type cleaning fluids (containing chlorine). The combined fumes can be deadly in any confined area. NEVER mix Clorox with ammonia-type cleaning fluids!

❖   ❖   ❖

PHOSPHORUS (Metalloids)
SYMBOL: P   **ATOMIC NUMBER:** 15   **PERIOD:** 3

COMMON VALENCE: 1, 3, 4, and 5 **ATOMIC WEIGHT:** 30.9738 **NATURAL STATE:** Solid **COMMON ISOTOPES:** There are no stable isotopes of phosphorus. Several man-made artificial isotopes are: phosphorus-29, phosphorus-30, phosphorus-31, phosphorus-32, phosphorus-33, and phosphorus-34. The most important is phosphorus-32. **PROPERTIES:** There are several **allotropes** of phosphorus (allotropy means "another manner," or the same element found in different forms). The allotropes of phosphorus are:

1. White phosphorus, a waxy, clear solid crystal. Density: 1.82. It must be stored under water to keep it from spontaneously igniting.

2. Red phosphorus, a violet-red powder derived from white phosphorus. Density: 2.34. It is not poisonous.

3. Black phosphorus, which looks something like graphite (a form of carbon). It is also derived from white phosphorus. Density: 2.5.

CHARACTERISTICS

White phosphorus occurs in nature in phosphate rock. It is **insoluble** in water and alcohol and will ignite spontaneously in air.

White phosphorus is **phosphorescent**, i.e., glows at room temperature. It must be stored under water.

Red phosphorus is less reactive than white phosphorus. Large amounts can explode. It is insoluble in most solvents.

Black phosphorus is insoluble in most solvents. It

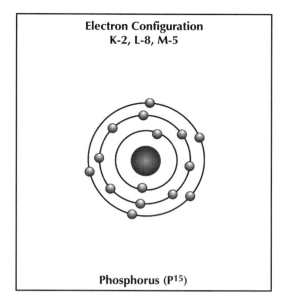

Electron Configuration
K-2, L-8, M-5

Phosphorus (P^{15})

will conduct electricity, while the white and red forms are poor conductors of electricity. It has little value.

ABUNDANCE AND SOURCE

Phosphorus is the eleventh most abundant element. It makes up about 0.1% of the Earth's crust.

Phosphorus occurs in nature in several forms, mostly as phosphates. The most common source is phosphate rock [$Ca_3(PO_4)_2$] and a mineral called "apatite." Phosphorus is found in all animal bones, teeth, and most living tissue. Phosphorus nodules are found on the ocean floor along with manganese nodules.

Most commercial phosphorus is produced in an electric furnace where phosphate-rich minerals are heated to drive off the phosphorus as a gas, which is then condensed under water. Another process uses sulfuric acid to remove the phosphorus.

HISTORY

In 1669, a German chemist, Hennig Brand (?–c.1692), was trying to create gold from urine. He did not get any gold, but after collecting urine in a bucket and allowing it to evaporate for several days, he then distilled it. From this residue he discovered a wax-like substance that glowed in the air. He named it *phosphorus* for the Greek word meaning "light bearer" after the bright early-morning star that had been given that name.

In 1841 Jöns Jakob Berzelius (1779–1848), who introduced the term "allotropy," transformed white phosphorus to red phosphorus. In 1865 Johann Wilhelm Hittorf (1824–1914) was the first to produce metallic phosphorus. Brand was given discovery credit.

COMMON USES

The allotropes and compounds of phosphorus have many important uses. It is an essential element in all living tissue, both plant and animal. Phosphorus is the main element in the compound adenosine triphosphate (ATP), which is the main energy source for living things.

White phosphorus is used to make wartime smoke screens, for rat and mice poisons, and as a **reagent** in chemistry.

Red phosphorus will ignite spontaneously. It is used to make safety matches, phosphoric acid, **electroluminescent** paints, and fertilizers.

Phosphorus has many other uses. Some examples are in pesticides, detergents, baking powder, ceramics, water softeners, and **pharmaceuticals**, and as an additive to both metals and petroleum products.

Black phosphorus is used in the electronics industry for its ability to conduct electricity.

EXAMPLES OF COMMON COMPOUNDS

Phosphorus-32, the most important radioisotope of phosphorus, has a half-life of 14 days, provides beta radiation (high-speed electrons), and is made by inserting phosphorus into nuclear reactor piles—Used as a "tag" to trace biochemical reactions in patients. Also used to treat leukemia, skin and thyroid diseases.

Phosphorus pentasulfide (phosphoric sulfide) (P_2S_5)—Used as an **insecticide**, as an additive to oils and to make safety matches.

Phosphorus trichloride (PCl_3)—Used to make other phosphorus compounds. Also used as an insecticide, as a gasoline additive, to make dyes, and to "finish" the surface of textiles.

HAZARDS

Many of the compounds of phosphorus are extremely dangerous, both as fire hazards and as deadly poisons to the nervous systems of humans and animals. Some of the poisonous compounds (PCL_x) can be absorbed by the skin as well as inhaled or ingested. Since some phosphorus compounds will continue to burn in water, they will continue to burn a deep hole if they come in contact with your skin.

Some of the main types of poisonous gases used in warfare, which are presently stockpiled in many countries, are phosphorus compounds.

❖ ❖ ❖

ARSENIC (Metalloids)
SYMBOL: As   **ATOMIC NUMBER:** 33   **PERIOD:** 4

COMMON VALENCE: 2, 3, and 5   **ATOMIC WEIGHT:** 75.92158   **NATURAL STATE:** Solid   **COMMON ISOTOPES:** There are no stable isotopes. Seventeen radioactive isotopes are known. **PROPERTIES:** Arsenic is considered a metalloid that has three allotropic forms of arsenic: a) the yellow cubic form and b) the black form, both of which revert to c) the stable silver-gray metallic form. Density: 5.72; melting point: 814°C; **sublimates** at temperatures between 620° and 814°C, which means it goes from a solid state directly to a gas without the intermediate stage of becoming a liquid.

CHARACTERISTICS

The elemental form of arsenic is a brittle, grayish crystal that becomes darker when exposed to air.

It is unique because, in its pure form, it does not melt, but changes from a solid directly to a gas (sublimation). It occurs in pure form in some minerals. There are few uses for arsenic.

ABUNDANCE AND SOURCE

Arsenic is the fifty-third most abundant element and is widely distributed in the Earth's crust. It occurs

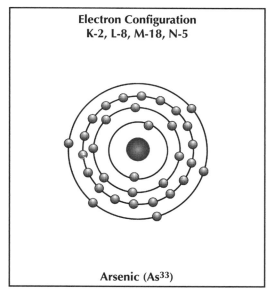

Electron Configuration
K-2, L-8, M-18, N-5

Arsenic (As³³)

naturally in several minerals, but high-grade deposits are rare. Most of the many minerals and ores that contain arsenic also contain other metals. Some major sources of arsenic are the minerals orpiment, scherbenkobalt, arsenopyrite, nicolite, realgar, gersdorffite, and smaltite. In addition, most sulfide ores contain arsenic. It is sometimes found pure, in uncombined form.

Arsenic is recovered as a by-product from the smelting of nickel, copper, cobalt, iron, and tin. It is also recovered from the flue dust of copper and lead smelting furnaces.

HISTORY

In about 3600 B.C. ores containing both arsenic and copper were known. This is about the time when copper was being smelted and alloyed to make bronze. Some ores of copper produced harder metals than others due to impurities. One of these impurities was arsenic. Since the workers became ill when smelting these types of ores, they were abandoned, and tin was added to the copper to form bronze. Bronze may have been the Persian word for "copper."

Early in the thirteenth century, Albertus Magnus (?–c.1280) proposed the concept of "affinitas," which explained how chemicals were held together. He was the first to figure out how to separate and identify the element arsenic.

COMMON USES

There are not many practical uses for arsenic.

It is used in the **semiconductor** industry to coat solid-state devices. Some compounds are used in paints and in fireworks. The major uses are in medicine, where its toxic properties are important for the treatment of some diseases.

At the beginning of the twentieth century arsenic compounds were used to kill spirochete bacteria, which cause the sexually transmitted disease syphilis. After the arsphenamine compound was used to treat syphilis in Europe, the disease rate was reduced by more than half. The antibiotic penicillin has replaced arsenic for most medical purposes.

Arsphenamine was called "606" because it was the 606th arsenic compound that Paul Ehrlich (1854–1915) had **synthesized** for use in treating diseases. His assistant found that it was effective as a treatment for syphilis. Ehrlich also coined the word *chemotherapy*. Historically, arsenic was one of the "poisons of choice" because it killed slowly and mimicked many other common ailments, such as **gastrointestinal** diseases, that could not easily be diagnosed and treated.

EXAMPLES OF COMMON COMPOUNDS

Arsenic disulfide (AsS)—Also known as ruby arsenic because it is a reddish-orange powder. Used as a **depilatory** agent, a paint pigment, and a rat poison; also used to make red glass and fireworks.

Arsenic trichloride (arsenic chloride) $(AsCl_3)$—Used in the **pharmaceutical** industry, and to make **insecticides** as well as ceramics.

Arsine (arsenic hydride) (AsH_3)—Used to synthesize organic compounds and as the major ingredient of several military poisons, including the wartime gas lewisite.

HAZARDS

The elemental forms of arsenic, as well as many of its compounds, are toxic when in contact with the skin, when inhaled, or when ingested. As with its cousin phosphorus, above it in Group 15 of the Periodic Table, care must be taken when using arsenic. Some compounds, such as arsenic trioxide (As_2O_3), are also **carcinogenic**. A poisonous dose as small as 60 milligrams can be detected within the body by the Marsh test.

❖   ❖   ❖

ANTIMONY (Metalloids)
SYMBOL: Sb   **ATOMIC NUMBER:** 51   **PERIOD:** 5

COMMON VALENCE: 3, 4, and 5 **ATOMIC WEIGHT:** 121.75 **NATURAL STATE:** Solid **COMMON ISOTOPES:** Antimony has 2 stable isotopes and several radioactive isotopes, including antimony-124, with a half-life of 60 days, and antimony-125, with a half-life of 2.4 years. Both emit **beta** and **gamma** radiation. **PROPERTIES:** Elemental antimony is a stable, bluish-white metal with a typical metallic luster. In addition, there are two **allotropes** of antimony (same element with different properties). They are yellow antimony crystals (Sb_4) and black amorphous antimony (Sb). Density: 6.69; melting point: 630.5°C; boiling point: 1380°C.

CHARACTERISTICS

Antimony and bismuth are the only elements in Group 15 that exhibit metal-like characteristics.

Both are hard, brittle metallics that have low melting points. They also exhibit poor heat and electrical conductivity.

Antimony is unique in that when it solidifies from a molten liquid state to a solid state, it expands, which is just the opposite of most metals.

ABUNDANCE AND SOURCE

Antimony is the sixty-third most abundant element on Earth, and it occurs

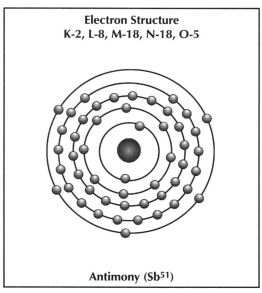

Electron Structure
K-2, L-8, M-18, N-18, O-5

Antimony (Sb^{51})

mainly as sulfide ores or in combinations with the ores of other metals. Sb_2O_3 represents several ores, e.g., stibnite, antimonite, and valentinite. It is seldom found free in nature.

Antimony is also found in copper, silver, and lead ores. Breithaupite (NiSb) and ullmannite (NiSbS) are two ores containing nickel. Dicrasite (Ag_2Sb) and pyragyrite (Ag_3SbS_3) are silver ores containing antimony.

HISTORY

Antimony was known in the days of alchemy (500 B.C. to A.D. 1600). It is possible that an alchemist, Basilius Valentinus, knew about antimony and some of its minerals and compounds around the middle of the fifteenth century. Even earlier than this period, physicians used metals, such as mercury and antimony, to try to cure diseases, even after they knew these elements were somewhat toxic.

Antimony was used to treat **depression**, as a laxative, and as an **emetic** for over two thousand years. The physicians of those days were aware that antimony was poisonous, but consided it a good medicine. Not much else was known about the chemistry of antimony until it was used to produce low-melting alloys.

COMMON USES

The most common use of antimony is as an alloy metal with lead to make the lead harder. This lead-antimony alloy is used for electrical storage batteries, sheathing electrical and TV cables, making wheel bearings, and **solder**.

It is also used in the **semiconductor** industry, for fireworks and to **vulcanize** rubber.

Both antimony-124 and antimony-125 radioisotopes are used as industrial and metallurgical **tracers**.

Some scholars believe that Mozart died from being given antimony by his physicians to treat his **depression**. They did not know how poisonous it was. The evidence for this story is scanty. It is also known that around 870 B.C., Queen Jezebel and her contemporaries used the mineral/ore antimony sulfide as a cosmetic to darken their eyelashes and as eyeliner. It is still used for this purpose in some countries.

EXAMPLES OF COMMON COMPOUNDS

Antimony-124, with a short, 60-day half-life—Used as a tracer to study solid-state materials. A unique use is as a partition or separator between different types of fluids flowing through the same pipelines. Its radioactivity can mark where one type of oil ends and another type of oil begins.

Antimony pentoxide (Sb_2O_3)—Used as a flame retardant for textile materials and as a source to prepare other antimony compounds.

Antimony trisulfide (Sb_2S_3)—Used as a yellow pigment in paint, in ruby glass, to make fireworks and matches. Also used to make percussion caps that set off explosives.

A unique use is to make glass that can reflect off **infrared** radiation; thus what is behind the glass cannot be detected by an infrared ray source.

HAZARDS

The powder and dust of antimony are toxic and can cause damage to the lungs. The fumes of antimony halogens (chlorides and fluorides) are especially dangerous when inhaled or in contact with the skin.

Many of the salts of antimony are **carcinogenic** and can cause lung cancer if inhaled, as well as other cancers if ingested. This is a major hazard with the radioisotopes of antimony used in industry. Some of the sulfides of antimony are explosive.

❖   ❖   ❖

BISMUTH (Metalloids)
SYMBOL: Bi   **ATOMIC NUMBER:** 83   **PERIOD:** 6

COMMON VALENCE: 3 and 5   **ATOMIC WEIGHT:** 208.980   **NATURAL STATE:** Solid   **COMMON ISOTOPES:** Bismuth-209 is the only stable isotope. There are 4 naturally occurring radioactive isotopes.   **PROPERTIES:** Bismuth is the most "metal-like" in its chemical and physical properties of the elements in Group 15. It is brittle, with a rosy-red lustrous tinge. Density: 9.8; melting point: 271°C; boiling point: 1560°C.

CHARACTERISTICS

Bismuth is more resistant to electrical current in its solid state than as a liquid.

Its thermal conductivity is the lowest of all metals, except mercury. Even though it is considered a metal-like element, it is a very poor conductor of heat and electricity.

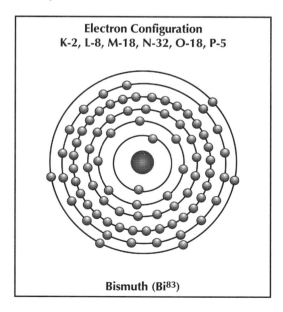

Electron Configuration
K-2, L-8, M-18, N-32, O-18, P-5

Bismuth (Bi[83])

ABUNDANCE AND SOURCE

Bismuth is the seventieth most abundant element, and it is widely spread over the Earth, but in very small amounts. It occurs in both the free elemental state and in several ores. The major ore, bismuthinite (Bi_2S_3), is found in South America.

The United States gets most of its bismuth as a by-product from smelting ores of lead, silver, copper, and gold. It is also recovered from refining tin and tungsten ores.

HISTORY

Bismuth, as well as arsenic, antimony, and zinc, were known as metals in the days of alchemy (500 B.C. to A.D. 1600). In the middle of the fifteenth century bismuth was described by a German monk, Basilius Valentinus (birth date unknown). He called it *wismut*, which is the German word for "white stuff." It was not identified as a separate element until 1659 when Johann Glauber used glass beads to demonstrate the importance of color in flames of different substances when trying to identify them. In 1753 Claud J. Geoffroy (birth date unknown) separated bismuth from its ore.

Not much is known of its use until the art of **metallurgy** developed in the late eighteenth century, when arsenic, antimony, and bismuth were used to form alloys with other metals.

COMMON USES

Bismuth is used to make drugs such as Pepto-Bismol for upset stomachs and diarrhea, and in medicine to treat intestinal infection. Bismuth is used in the cosmetic industry to provide the "shine" for lipstick, eyeshadow, and other cosmetics.

Bismuth is added to steel to make it easier to roll, press, pull into wires, and turn on a **lathe**.

It is also used in the **semiconductor** industry and to make permanent magnets.

Bismuth is similar to antimony in that it expands from the molten liquid state to the solid state. This property makes it an excellent material to make molds, which can be used to produce fine details in whatever is being molded. It is used as molds to produce metallic printing type and similar fine castings.

EXAMPLES OF COMMON COMPOUNDS

Bismuth antimonide (BiSb), single crystals of the alloy of bismuth and antimony—Used as semiconductors in the electronics industry. Also used to produce type for printing presses and low-melting-point electrical fuses.

Bismuth subcarbonate [(BiO)$_2$CO$_3$]—Used to make other bismuth compounds, cosmetics, enamel, and ceramic glazes. Its major use is as an opaque substance placed in the digestive tract to show up on X-rays. X-rays are blocked by the bismuth; thus the physician can see patterns inside the stomach or intestines.

Bismuth subnitrate (BiNO$_3$)—Used in the cosmetics industry, and to make ceramic and enamel glazes.

HAZARDS

Bismuth is flammable as a powder. The halogen compounds of bismuth (Group 17) are toxic when inhaled or ingested. Some of the salts of bismuth can cause metallic poisoning similar to mercury and lead.

At the beginning of the twentieth century, before penicillin, bismuth compounds were used to treat some venereal diseases. However, the treatment did not work very well.

Group 16 Elements
THE NONMETALS

The elements in Group 16 have characteristics ranging from metalloids or semiconductors to nonmetals. The first two, oxygen (O^8) and sulfur (S^{16}), are quite reactive nonmetals and readily accept 2 electrons from metals and other elements to complete their outer shells, forming negative ions. Or, they can share electrons to complete the required 8 electrons in their outer shells. Thus, they are strong **oxidizers**.

Several of the elements in Group 16, e.g., Se^{34} and Te^{52}, exhibit **semiconductor** characteristics. The heaviest element in Group 16, polonium (Po^{84}), is a radioactive poison.

Since the elements in Group 16 all have 6 electrons in their outer orbits, they tend to either "collect" electrons from metals to form **ionic bonds** or share electrons to form **covalent bonds** with other elements. They are: O^8, S^{16}, Se^{34}, Te^{52}, and Po^{84}.

OXYGEN (Nonmetals)
SYMBOL: O   **ATOMIC NUMBER:** 8   **PERIOD:** 2

COMMON VALENCE: 2 **ATOMIC WEIGHT:** 15.9994 **NATURAL STATE:** Gas **COMMON ISOTOPES:** There are 3 stable nonradioactive isotopes of oxygen: oxygen-16, oxygen-17, and oxygen-18. **PROPERTIES:** Oxygen is a very reactive nonmetal. There are three allotropes (different forms) of oxygen: 1) atomic oxygen (O), sometimes referred to as nascent or "newborn" oxygen; 2) diatomic oxygen (O_2), or molecular oxygen (gas); and 3) ozone (O_3). Oxygen gas is colorless, odorless, and tasteless. Oxygen readily forms compounds with most other elements. Since it has 6 electrons in its outer valence shell, it easily gains 2 electrons to form a negative ion 2– or shares electrons to complete its outer shell.

Almost all of the oxygen in the atmosphere is made up of the **diatomic** molecule O_2. This essential gas we breathe is the result of **photosynthesis**, which is how green plants (with chlorophyll) use the energy of the sun to convert carbon dioxide (CO_2) and water to starches/sugars with molecular oxygen (O_2) as a by-product. Density: 1.29 (as a gas); density: 1.14 (as a liquid). Oxygen gas liquefies at –183°C and turns to a solid at –218°C.

CHARACTERISTICS

Oxygen is, without a doubt, the most essential element on Earth. It is required to support all plant and animal life.

Oxygen is soluble in both water and alcohol.

Contrary to what many people think, oxygen is NOT combustible (it will not burn), but rather it actively supports the combustion of many other substances. After all, if oxygen burned, every time you lit a fire all the O_2 in the atmosphere would be consumed!

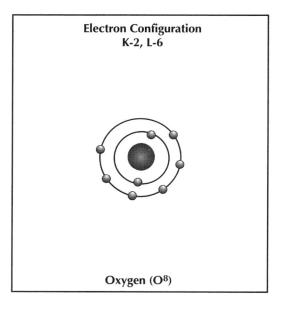

Electron Configuration
K-2, L-6

Oxygen (O^8)

Burning is a form of **oxidation** where oxygen chemically combines with a substance rapidly enough to produce adequate heat to cause fire and light, or to maintain a fire once started.

The oxidation of iron is called rusting. Rusting is an example of "slow oxidation," which is the reaction of O_2 with Fe to form Fe_2O_3 or Fe_3O_4. This **chemical reaction** is so slow that the heat it does produce is dissipated; thus, there is no fire.

ABUNDANCE AND SOURCE

Oxygen is the most abundant element in the Earth's crust, in water, and in the atmosphere. Combined with other elements, oxygen makes up almost one-half of the total mass of the Earth's crust. However, oxygen gas (O_2) makes up about 21%, or about 1/5, of the Earth's atmosphere by volume.

Oxygen will combine with almost any other element—even some of the inert noble gases, which do not easily combine with anything.

Oxygen can be purified by liquefying air and then fractionating the liquid air to separate the oxygen from the nitrogen and trace gases. **Fractionation** is the process of separating and isolating a mixture of either gases or liquids. It is based on the principle that each element has its own temperature at which it turns from a liquid to a gas. By "boiling off" the liquid oxygen at the temperature at which it changes from a liquid to a gas, it can be separated from the other liquefied components of air and collected as pure oxygen gas. This is possible since each element in liquid air (mainly nitrogen) has a specific boiling point, which is different from that of oxygen.

A more recent, and productive, method is to pass air through fine molecular-size sieves of material that will absorb the nitrogen gas, which allows the oxygen gas to pass through the sieve to be collected.

HISTORY

Historically, several scientists were working with the concept that air consists of more than one gas. They all knew that something in air supported life. Joseph Priestley (1733–1804) gave his new gas the name *dephlogisticated air*, Carl Wilhelm Scheele (1742–1786) called his new gas *empyreal air*, and Daniel Rutherford (1749–1819) called the gas he worked with *phlogisticated air*.

Antoine-Laurent Lavoisier (1743–1794) followed up Priestley's work by making quantitative measurements of the ratio of oxygen to nitrogen in air. At first he named the new gas *highly respirable air*, and later, *vital air*. Lavoisier is often considered the "father of modern chemistry" for the experimental procedures he used and for making precise measurements and recording his data when conducting his investigations. Credit for naming the element was given to Lavoisier. He and other scientists of his time believed that all acids contained oxygen. Because acids smell "sharp," Lavoisier named the element *oxygen*, from *oxys*, the Greek word for "sharp." Later it was discovered that not all acids contain oxygen, e.g., hydrochloric acid (HCl).

COMMON USES

Besides the essential use to support life, oxygen has many other uses.

It is used in the **smelting** process to free metals from their ores. It is particularly important in the oxygen converter process when making steel from iron ore.

Oxygen is used in making several important **synthetic** gases and in the production of ammonia, methyl alcohol, etc.

It is the oxidizer for liquid rocket fuels, and as a gas, oxygen is used in a mixture with helium to support the breathing of astronauts and divers and to aid patients who have difficulty breathing.

It is used to treat (oxidize) sewage and industrial organic wastes.

Oxygen has many uses due to its ability to accept electrons from other elements to form **ionic bonds**, or to share electrons with other elements to form **covalent bonds**.

As previously mentioned, oxygen is the by-product of **photosynthesis**, where carbon dioxide and water are transformed into food by the chlorophyll of green plants, using sunlight as energy. Carbon dioxide is produced by the decay of organic materials and the burning of **fossil fuels** in homes, in industry, and for transportation. There is much concern about the increase of carbon dioxide in our atmosphere. But at the same time, an atmosphere enriched with CO_2 enhances the growth of plants, which are ultimately the source of all our food.

The respiration (breathing) of plants and animals also involves the intake of oxygen, followed by the process of oxidation, which converts food into energy that is required to sustain all life.

EXAMPLES OF COMMON COMPOUNDS

Oxygen forms many compounds. It combines directly with single elements or with several other elements to form a large number of oxygen compounds, referred to as **oxides**. The most important binary molecular compound is water, H_2O.

Silicon dioxide (SiO_2), common sand, is one of the most plentiful compounds on Earth.

There are numerous oxygen compounds containing more than two elements. They compose the silicates, which make up rocks and soil, as well as limestone (calcium carbonate), gypsum (calcium sulfate), bauxite (aluminum oxide), and many iron oxides.

HAZARDS

While oxygen itself is not flammable or explosive, as is sometimes believed, its main hazard is that in high concentrations oxygen gas can cause other things to burn much more rapidly (rapid oxidation or combustion). This is what happened in one of NASA's space capsules before it was launched. The level of oxygen of the interior atmosphere was increased to the point that when a small fire started, which would have been minor in a normal air atmosphere, it became a raging inferno. This caused the deaths of the astronauts inside the capsule.

Oxygen used for therapeutic purposes in adults can cause **convulsions** if the concentration is too high. At one time, high levels of oxygen were given to premature infants to assist their breathing. It was soon discovered that a high concentration of O_2 caused blindness in some of the infants. The level has been reduced, and this is no longer a medical problem.

OZONE

PROPERTIES

Ozone is an allotropic molecular form of oxygen containing three atoms (O_3). It is a much more powerful oxidizing agent than diatomic oxygen (O_2) or monatomic oxygen (O). Instead of being colorless, like oxygen gas, ozone is bluish in the gaseous state, but blackish-blue in the liquid and solid states (similar in color to ink). Ozone's boiling point is $-112°C$, and its freezing point is $-192°C$.

CHARACTERISTICS

Ozone has a very distinctive pungent odor. It exists in our lower atmosphere in very small trace amounts. In high concentrations it is irritating and even poisonous.

Electrical discharges in the atmosphere produce ozone. This is what gives the air a "clean" smell after lightning, and you can also recognize the odor when running electrical equipment that gives off sparks. Even toy electric trains can produce ozone as they spark along the track. Ozone can be produced by passing dry air between two electrodes that are connected to an alternating electric current with a high voltage.

Besides being produced by electrical discharges, ozone is produced in the upper atmosphere or stratosphere by **ultraviolet** (UV) light from the sun striking O_2 molecules, breaking them down and reforming them as O_3 molecules. The vast majority of ozone is produced in the atmosphere over the tropical latitudes because this area gets the most sunlight. Normal wind currents carry the ozone to the polar regions of the Earth where it is thickest.

HISTORY

In 1839 Christian Friedrich Schonbein (1799–1868) discovered a gas with an unusual odor coming from some electrical equipment. He did not know what it was, but since it had an odd smell he called it *ozone*, after the

Greek word for "I smell." Although he knew that it was a chemical substance, he mistakenly associated ozone with the halogens (Group 17). Others before Schonbein had smelled the gas but did not recognize its importance. Thomas Andrews (1813–1885) and several other scientists, through different experiments, identified ozone as a form of oxygen (allotropy). It was not until 1868 that J. Louis Soret established the formula to be O_3.

USES

Ozone, produced by electrical discharges, is used to purify drinking water and to treat industrial wastes and sewage. It is also used to deodorize air and kill bacteria by passing air through special ozone-producing electronic devices.

Because of its powerful oxidizing characteristics, it is useful in several industrial processes, such as: the manufacture of paper, steroid hormones, waxes, and cyanide, and the processing of acids.

HAZARDS

High concentrations of ozone are a fire and explosion hazard when in contact with any organic substance that can be oxidized.

In high concentrations ozone is very toxic when inhaled and irritating to the nose and eyes in lesser concentrations. The Environmental Protection Agency (EPA) standard for ozone in air of the lower atmosphere is only 0.12 ppm.

Ozone in the lower atmosphere contributes to air pollution and smog. It can cause damage to rubber, plastic, and paints. These low concentrations can cause headaches, burning eyes, and respiratory irritation. It is particularly harmful to asthmatics and the elderly with respiratory problems.

OZONE CONTROVERSY

The effects of natural and man-made chemicals on the ozone layer of the upper atmosphere have been suspected and controversial for over thirty years. In 1974 F. Sherwood Rowland and Mario Molina, at the University of California at Irvine, suggested that the problem chemicals were **chlorofluorocarbons** (CFCs) used in refrigeration and as a propellant in spray cans. Several chemists suspected the effects of CFCs on the ozone layer as early as the 1960s. They had experimental laboratory evidence of the **chemical reactions** involved, but no actual data that related to the ozone layer. Since the mid-1970s, both the scientific facts and the arguments about potential problems have become ensnared in political and environmental ideologies.

The ozone issue, as well as the possibly related issue of global warming, as well as other environmental concerns, is often related to a lack of understanding of both the science and the political arguments involved.

In 1913 Charles Fabry (1867–1945) identified large quantities of ozone in the upper atmosphere at about 10 to 30 miles. This layer is called the **ozonosphere**, or more commonly, the ozone layer.

The ozone gas layer of the stratosphere, although thin, is massive and covers most of the Earth at high altitudes. It is essential because it partially blocks and absorbs the UV rays of the sun, preventing them from reaching the Earth, particularly the tropical and subtropical areas. Not all the UV is absorbed by the ozone, but a large portion of the stronger, shorter-wave-length rays, which are the most harmful to living organisms, are absorbed and do not reach the Earth. Some UV **radiation** is needed to activate vitamin D, but excessive ultraviolet light can cause skin cancer, cataracts, mutations, and death in plant and animal cells. One theory is that early in the formation of life and in the following ages, the lack of an ozone layer permitted short UV rays to reach Earth, and thus may have been partially responsible for the proliferation of species due to genetic **mutations**.

Through the natural process of the UV rays of the sun passing through this layer, the O_3 absorbs the rays and is broken down to O_2 molecules and O atoms. This process is reversible and ozone (O_3) is constantly being reformed from UV effects on O_2. However, the separation can be accelerated faster than the reformation of new O_3 by the induction of other chemical gases into the ozone layer. Of particular concern is that chlorine from CFCs, and from other sources such as the ocean, combines with atomic oxygen that is broken down from O_3 by UV radiation. It then forms chlorine monoxide (ClO), which means the atomic oxygen is not available for reformation into O_3. Herein lies the potential problem and the controversy.

At issue is a group of man-made chemicals called chlorofluorocarbons (CFCs), which are used as the refrigeration fluids (Freon) in automobile, home, and industrial air-conditioning and refrigeration units and which, in the past, were used as the pressure gas in spray cans of paint, deodorant, and so forth.

When CFCs slowly rise in the atmosphere and reach the ozone layer, they are broken down into component molecular compounds and atoms by the UV rays of the sun. Some of these chemicals then react with ozone to break it down, thus reducing the amount of O_3. Further, chlorine and some other elements combine with the O and O_3 to form other chemicals. This also contributes to the reduction of ozone faster than it can be reformed by the natural processes. Ozone is a renewable resource. The issue is, can a balance be obtained between the destruction of ozone in the atmosphere, by both natural and man-made causes, and its natural regeneration?

Also of concern because of their possible connection with global warming are hydrocarbon gases such as methane, which is produced naturally, in large quantities, as the gas from the digestive process of cows, from decaying organic matter, and from petroleum refining, and the gas carbon dioxide, which is produced in nature by respiration of plants and animals, by volcanic action and forest fires, and by humans in the burning of wood, coal, gas, and oil. Nitrogen-based compounds from automobile exhausts also contribute to the problem, but ozone is not directly produced by automobile exhaust gases.

There are parallel policy issues for global warming and the periodic increase and decrease of the ozone layer, but the science is not parallel. The following are some things to keep in mind when examining these issues:

1. There is agreement that a "hole" in the ozone layer over Antarctica does exist, and that it changes size over periods of time. But, not all scientists agree on the associated causes, seriousness, dangers, or cures.

2. The ozone layer is dynamic and unpredictable, which means it is constantly changing and seems to change in ways we cannot yet understand. The large hole found in the ozone layer over Antarctica seems not only to move but to get larger, then smaller. The ozone layer is thickest over the poles of the Earth, yet it is mostly produced over the equator.

3. Much of the data supporting claims on both sides of the controversy of ozone depletion and global warming need to be analyzed with great care, since the issues have gone beyond the science involved. Our ability to predict long-term weather and atmospheric conditions, even with computer modeling, is limited. A new science called "chaoplexity," which combines chaos theory with related problems of complexity, may lead to a useful method of predicting "unpredictable" events. However, it may not be useful.

4. Global warming may or may not be a related problem to ozone depletion. There has been a 30% increase in the amount of carbon dioxide produced by humans since the early 1700s. Carbon dioxide makes up just 3% of the atmosphere. Only a small fraction of this CO_2 has remained in the atmosphere. Some is used by plants to make food (**photosynthesis**), some is dissolved in the oceans, and some combines with other elements. The argument is that CO_2 and some other gases from the burning of hydrocarbon fuels may form a "greenhouse" effect in the upper atmosphere to hold in the heat of the Earth and/or have some effect on the ozone layer.

According to the Council on Environmental Quality, there was a warming trend of the Earth from 1870 to 1940. This trend was reversed from 1940 to 1960 when the Earth became cooler. The hottest year on record was 1995, which was just 0.8°F above the former record set in 1990. There has been a general (average) temperature increase of about 1°F of the Earth's atmosphere over the last 100 years. Some of this increase is due to natural atmospheric conditions and some to increased concentrations of trace gases, such as carbon dioxide, methane, CFCs, sulfur, and nitrogen compounds. The good news is that there now seems to be a reversal of this warming trend due to a cooling of the water in the tropical Pacific Ocean. Many scientists feel that part of the problem is our limitations in making accurate long-term measurements and predictions. Another problem is that scientists and the media do not treat the same data in similar ways. Their ideological agendas may be different, which can affect how data are presented to the public.

Today, there are over 30% more trees growing in the United States than there were 100 years ago. Sadly, this is not true for all countries, some of which have depleted not only their forests but most natural vegetation, causing the formation of deserts. One way of reducing the amount of carbon dioxide in the atmosphere, besides reducing the burning of hydrocarbon fuels, is to greatly increase plant life on Earth. We need to reestablish forests worldwide.

5. The U.S. government banned the use of chlorofluorocarbon aerosols in paint and spray cans in 1978. In 1986 an international agreement was adopted that requires all industries to reduce the manufacture and use of CFCs by 50% by the year 2000. More recently, an outright ban on the use of Freon (CFCs) by industrialized countries has been agreed to. Research is being conducted to find suitable substitutes. One group of promising compounds are called HFCs and HCFCs, which break down before reaching the ozone level of the stratosphere. The problem is that some of these new HCFCs are flammable and possibly **carcinogenic**.

6. It is doubtful if any international agreement to control the so-called greenhouse gases by all nations can be effective. First, many developing nations will continue to develop industries that produce these gases. Second, the cost of developing alternate, nonpolluting energy sources will be tremendous. And, finally, Mother Earth, through many natural processes, will continue to produce variable amounts of gases, such as carbon dioxide, chlorine, methane, oxygen, and ozone, in the future and tend to establish her own balances, as she always has.

Much more scientific study and the objective generation and interpretation of reliable data are warranted.

❖   ❖   ❖

SULFUR (Nonmetals)
SYMBOL: S   ATOMIC NUMBER: 16   PERIOD: 3

COMMON VALENCE: 2, 4, and 6 **ATOMIC WEIGHT:** 32.064 **NATURAL STATE:** Solid **COMMON ISOTOPES:** Sulfur has 4 stable isotopes: sulfur-32, sulfur-33, sulfur-34, and sulfur-36. Sulfur-32 makes up over 95% of all sulfur found on Earth. Sulfur-35 is a man-made radioisotope with a half-life of 87.1 days. It is made in **nuclear reactors** (piles). **PROPERTIES:** Sulfur is considered a nonmetallic solid. It is found in two **allotropic** crystal forms:

1. Rhombic, octahedral lemon-yellow crystals, which are also called "brimstone" and also referred to as "alpha" sulfur. Its density is 2.06 with a melting point of 95.5°C

2. Monoclinic, prismatic crystals, which are light-yellow. It is referred to as "beta" sulfur. Its density is 1.96 with a melting point of 119.3°C.

There are two liquid amorphous forms of sulfur.

CHARACTERISTICS

When sulfur is melted, it exhibits unique properties. It increases its **viscosity** and turns reddish-black as it is heated. Beyond 200°C the color then lightens and it flows as a thinner liquid.

Above 445°C sulfur turns to gas, which is dark orangish-yellow but becomes lighter in color as the temperature rises. Sulfur has the ability to combine with most other elements—except the inert gases—to form compounds.

ABUNDANCE AND SOURCE

Sulfur is the seventeenth most abundant element and makes up about 1% of the Earth's crust. It is not unusual to find sulfur as a free element in nature, especially in the vicinity of volcanos.

Sulfur is mined by the Frasch process, which was invented by Herman Frasch

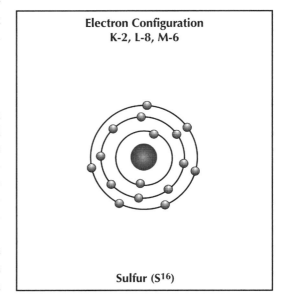

Electron Configuration
K-2, L-8, M-6

Sulfur (S^{16})

in the early 1900s. This process forces superheated water, under pressure, into deep underground sulfur deposits. Compressed air then forces the molten sulfur to the surface, where it is cooled. There are other ways to mine sulfur, but the Frasch process is the most important and the most economical.

Sulfur is found in Sicily, Canada, Central Europe, and the Arabian oil states, as well as in Texas, Louisiana, and offshore beneath the Gulf of Mexico.

HISTORY

Sulfur was known to early humans. It was possibly one of the first free elements humans tried to use and understand besides the "always known" air, fire, and water.

Sometime around A.D. 1300 an unknown alchemist described sulfuric acid. We don't know much about the early use of sulfur or sulfuric acid. In 1579 an alchemist named Andreas Libavius described the progress of alchemy. In his book he described how hydrochloric and sulfuric acids are produced. He also mentioned the formation of **aqua regia,** which is a mixture of acids that is strong enough to dissolve gold—the royal metal.

In 1777 Antoine-Laurent Lavoisier (1743–1794), the "father of modern chemistry," was the first to recognize sulfur as an element.

COMMON USES

Sulfur is a very important element for today's industries and for the well-being of all of us.

Most sulfur that is mined is used to produce other chemicals. Large amounts of sulfur or sulfur compounds are used in industries that produce the following items: matches, gunpowder, soaps, fertilizer, plastics, bleaching agents, rubber, road asphalt, binders, **insecticides**, paints, dyes, medical ointments, other **pharmaceutical** products, and many more.

Sulfur is essential to life.

EXAMPLES OF COMMON COMPOUNDS

Hydrogen sulfide (H_2S) is one of the most important compounds. It is the colorless gas that has the foul, rotten-egg odor. It is used in the production of both sulfur and hydrogen, as well as sulfuric acid.

There are so many metal sulfides that chemists usually use the letter "M" in the formula to indicate that sulfur can combine with just about any metal, e.g., MHS, M_2S, M_2S_3, etc.

Sulfuric acid (H_2SO_4), also known as battery acid, is by far the most important and most commonly used industrial chemical in the world. Over 80 billion pounds are used in the United States each year. It is strongly corrosive and, in both concentrated and weak solution with water, will attack most metals.

Sulfur chloride (S_2Cl_2) is combustible and will react when in contact with water. It is used to produce carbon tetrachloride, to purify sugar juices, in insecticides, to extract gold from its ore, and as a military poisonous gas.

There are 4 major phosphorus-sulfur compounds, ranging from P_4S_3 to P_4S_{10}, and there are 6 important oxygen-sulfur compounds, ranging from SO to S_2O_7, the most important being sulfur dioxide (SO_2), which can act as both an **oxidizing** and a **reducing agent**.

HAZARDS

Many of the sulfur compounds are toxic but essential for life. The gas from elementary sulfur and from most of the compounds of sulfur is poisonous when inhaled and deadly when ingested. This is what makes some sulfur compounds useful for killing rats and mice.

Sulfa drugs (sulfanilamide and sulfadiazine), although toxic, were used as medical antibiotics before the development of penicillin. They are still used today in veterinary medicine.

❖   ❖   ❖

SELENIUM (Nonmetals)
SYMBOL: Se   **ATOMIC NUMBER:** 34   **PERIOD:** 4

COMMON VALENCE: 2, 4, and 6   **ATOMIC WEIGHT** 78.96   **NATURAL STATE:** Solid   **COMMON ISOTOPES:** There are 6 stable isotopes of selenium. **PROPERTIES:** Selenium has many of the same properties as tellurium, which is just below it in Group 16 of the Periodic Table. Both are considered metalloids (semiconductors). In the amorphous state it is a red powder that turns black and becomes crystalline when heated. Crystalline selenium has a density of 4.5, a melting point of 217°C, and a boiling point of 685°C.

CHARACTERISTICS

Crystalline selenium is a p-type **semiconductor**.

It acts as a **rectifier** and has **photovoltaic** properties, which means it is able to convert light (radiant) energy into electrical energy.

Selenium burns with a blue flame that produces selenium dioxide (SeO_2). It will react with many metals as well as with nonmetals, including the halogens (Group 17).

ABUNDANCE AND SOURCE

Selenium is the sixty-seventh most abundant element. It is widely spread over the Earth, but does not exist in large quantities. As a free element, it is found with elemental sulfur.

The most important commercial source is found in selenium-copper ores, from which it is extracted in usable amounts.

Selenium is found in Mexico, Bosnia, Japan, and Canada. It can be found in recoverable quantities in some soils.

HISTORY

In the early 1800s Jöns Jakob Berzelius (1779–1848), along with his friend Wilhelm Hisinger (1766–1852), who was a mineralogist, discovered the element cerium, which they named after the just discovered asteroid *Ceres*. Another element they discovered was named *selenium*, after the Greek word *selene*, which means "moon."

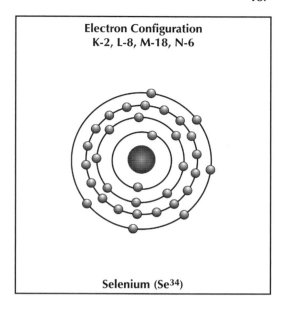

**Electron Configuration
K-2, L-8, M-18, N-6**

Selenium (Se^{34})

COMMON USES

Selenium's use as a semiconductor is not as important as it was some years ago. The elements silicon and germanium are more often used now for this purpose.

Selenium still finds uses in the electronics industry to make TV cameras, solar cells, light exposure meters for photography, xerography, and so forth.

It is also used to color ceramics, to produce copper and steel, to make inks, and as a catalyst (to speed up the reaction) for making rubber.

Selenium is an essential trace element for both plants and animals. It is used as a diet supplement in animal feeds as well as for humans.

EXAMPLES OF COMMON COMPOUNDS

Selenium dioxide (SeO_2)—Used as an **oxidizing agent**, as a catalyst, and as an **antioxidant** for lubricating oils and grease.

Selenium sulfide (SeS_2)—Used for some medicines, as an additive for medicated shampoos to control dandruff and scalp itching, and in treatment products for acne and eczema.

HAZARDS

The fumes and gases of most selenium compounds are very toxic when inhaled. SeO_2 and SeS_2 are toxic if ingested and very irritating to the skin. They are also **carcinogenic**.

Although some compounds of selenium are poisonous, as an element it is essential in trace amounts for humans. It is recommended that 1.5 to 5 milligrams of selenium be included in your daily diet. This amount can be maintained by eating seafood, egg yokes, chicken, milk, and whole-grain cereals. Selenium assists vitamin E in preventing the breakdown of cells and some chemicals in your body.

❖ ❖ ❖

TELLURIUM (Nonmetals)
SYMBOL: Te   ATOMIC NUMBER: 52   PERIOD: 5

COMMON VALENCE: 2, 4, and 6 **ATOMIC WEIGHT:** 127.60 **NATURAL STATE:** Solid **COMMON ISOTOPES:** There are 8 stable isotopes of the element tellurium. **PROPERTIES:** Tellurium is a silver-white solid with a metallic luster and exhibits metalloid (**semiconductor**) characteristics. It is similar to selenium and sulfur just above it in Group 16 of the Periodic Table.

There are two **allotropic** forms of tellurium: 1) The crystalline form, which has a silver-white metallic appearance with a density of 6.24, melting point of 450°C, and boiling point of 1000°C, and 2) the **amorphous** form, which is brown with a density of 6.015 and similar ranges for the melting and boiling points.

CHARACTERISTICS

The pure form of tellurium burns with a blue flame and forms tellurium dioxide (TeO_2).

It will react with the halogens of Group 17, but not with Se or S of its own group.

Tellurium is soluble in acids but not in water.

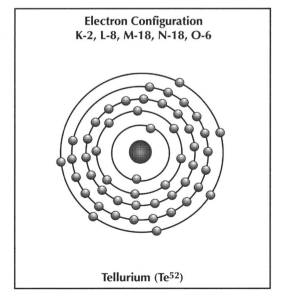

Electron Configuration
K-2, L-8, M-18, N-18, O-6

Tellurium (Te^{52})

ABUNDANCE AND SOURCE

Tellurium is the seventy-second most abundant element on Earth. It makes up a very small portion of **igneous rocks**. It is sometimes found as a free element, but more often it is recovered from ores of other elements, such as sylvanite (also known as graphic tellurium), nagyagite, black tellurium, hessite, tetradymite, altaite, coloradoite, and silver and gold tellurides, and is sometimes found along with selenium.

Tellurium is also recovered from the "slime" that collects on the **anode** in the **electrolyte** used to refine copper and lead.

HISTORY

Two people are credited with the discovery of tellurium. First, Franz Joseph Muller (1740–1825), who was working with gold ore, found something he thought might be a new element. He sent a sample to Martin Heinrich Klaproth (1743–1817) who correctly identified it as a new element. It was named *tellurium* for the Latin word meaning "Earth."

COMMON USES

The most common use of tellurium is to produce alloys of iron and steel. The alloy, sometimes along with other metals, makes stainless steel and casting brighter and easier to work with.

It is also used as a **vulcanizing** agent when making rubber, as a coloring agent for glass and ceramics, and for **thermoelectrical** devices.

Along with lithium, it is used to make special batteries for spacecraft and **infrared** lamps.

Tellurium can be used as a p-type semiconductor, but more efficient elements can do a better job of converting light energy to electrical energy.

It is also used as a **depilatory**, which removes hair from skin.

EXAMPLES OF COMMON COMPOUNDS

Tellurium dibromide ($TeBr_2$)—Forms blackish-green needles that are very **hygroscopic** (readily absorbs water). Toxic when inhaled.

Tellurium dichloride ($TeCl_2$)—Similar to $TeBr_2$, but in powder form it is greenish-yellow. It is also toxic when inhaled.

Tellurium dioxide (TeO_2)—A whitish crystalline powder that is slightly soluble in water. It is also toxic when inhaled.

HAZARDS

All forms of tellurium are toxic in gas form. The vapors of all the compounds or the dust/powder of the element tellurium should not be inhaled or ingested.

When a person is poisoned with tellurium, even in a small amount, his or her breath smells like garlic.

❖   ❖   ❖

POLONIUM (Nonmetals)
SYMBOL: Po   ATOMIC NUMBER: 84   PERIOD: 6

COMMON VALENCE: 2, 4, and 6 **ATOMIC WEIGHT:** 210 **NATURAL STATE:** Solid **COMMON ISOTOPES:** There are no stable isotopes of polonium. There are at least 34 unstable radioactive isotopes of polonium, more than for any other element. They range from polonium-192 to polonium-218, and beyond. Polonium-210 is the only natural radioactive isotope with a half-life of 138.5 days. Two important man-made radioisotopes are polonium-208 with a half-life of 2.9 years and polonium-209 with a half-life of 100 years. **PROPERTIES:** Most of the known chemistry of polonium is based on the naturally occurring radioactive isotope polonium-110, which is a natural radioactive decay by-product of the uranium family. Density: 9.4; melting point: 254°C; boiling point: 962°C.

CHARACTERISTICS

Polonium is more metallic in its properties than the elements above it in Group 16 of the Periodic Table, and it is the only element in this Group that is naturally radioactive.

It is chemically related and similar in reddish color to tellurium when forming oxides. Polonium is soft, resembling thallium, lead, and bismuth.

ABUNDANCE AND SOURCE

Polonium is found only in trace amounts in the Earth's crust. In nature, it is found in **pitchblende** as a decay product of uranium. Radioactive isotopes of polonium can be produced by bombarding bismuth with neutrons in a nuclear (atomic) reactor.

HISTORY

In 1898 Marie Sklodowska Curie (1867–1934) had identified several other radioactive elements during her work with pitchblende,

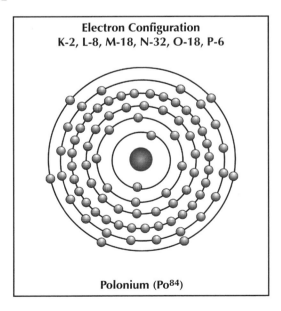

Electron Configuration
K-2, L-8, M-18, N-32, O-18, P-6

Polonium (Po⁸⁴)

the ore of uranium. Her husband, Pierre Curie (1859–1906), had discovered **piezoelectricity**, which could be used to measure the strength of radiation given off by the radioactive elements his wife was working with.

Madame Curie is credited with discovering not only uranium and radium, but also polonium, which she named after her native country, Poland. She is also credited with coining the word *radioactivity*.

Madame Curie is one of only two chemists to receive two Nobel Prizes. In 1903, both the Curies and Antoine-Henri Becquerel (1852–1908) shared the Prize for discovering radium, and in 1911 Madame Curie received the Prize for chemistry. The other was Linus Pauling (1901–1994), who received one Nobel Prize for chemistry in 1954, and a Nobel Peace Prize in 1962. Madame Curie died from radiation poisoning she received from working with radioactive elements.

COMMON USES

There are not many common uses for polonium. Probably the most important is as a source of **alpha particles** (nuclei of helium atoms) and high-energy neutrons for radiation studies. It is also used to calibrate radiation-detection instruments.

It is used to make a special alloy for spark plugs that make starting your car easier in cold weather.

It can be used as a portable low-level radioactive power source.

EXAMPLES OF COMMON COMPOUNDS

Both $SPoO_3$ and $SePoO_3$ are bright red in color, similar to tellurium compounds.

Polonium will form compounds with oxygen and some metals, but they are not common.

HAZARDS

Even though polonium is a rare element, it is a very dangerous radiation hazard and should be avoided.

Cigarette smoke contains polonium, which, along with many other **carcinogenic** chemicals, causes lung cancer. Over one hundred elements and compounds have been identified in cigarette smoke besides polonium. Some examples are nicotine, cresol, carbon monoxide, pyridene, and the **carcinogenic** compound benzopyrene.

8

HALOGENS AND NOBLE GASES

Group 17 Elements
THE HALOGENS

The halogens are the five chemical elements in Group 17 (VIIA) on the right side of the Periodic Table. They are fluorine, chlorine, bromine, iodine, and astatine.

Their unique characteristics are due to their outer shells having 7 electrons, and thus requiring only 1 electron to become complete. This makes the halogens very reactive with both metals and some nonmetals. They are all considered **monovalent**, poisonous, nonmetallic elements that form negative ions and may form either **ionic** or **covalent** bonds. They can also form compounds with each other and can form **halides**.

When they form ionic bonds, they gain 1 electron in their outer shells to form monovalent negatively charged ions, which are known as "halide ions." You might think of the halogen ions as now having the same electron structures as the inert noble gases of Group 18 (VIIIA), except that they have a negative charge.

Halogens are excellent oxidizing agents. As their atoms become larger, i.e., as you go down in the Group, they are less effective as oxidizing agents. Fluorine, at the top of Group 17, is the least dense and the most reactive, while iodine, near the bottom, is the least active (excluding astatine, which

PERIODIC TABLE OF THE ELEMENTS

TRANSITION ELEMENTS

GROUPS / PERIODS	1 IA	2 IIA	3 IIIB	4 IVB	5 VB	6 VIB	7 VIIB	8 VIII	9 VIII	10 VIII	11 IB	12 IIB	13 IIIA	14 IVA	15 VA	16 VIA	17 VIIA	18 VIIIA
1	1 H 1.0079																	2 He 4.00260
2	3 Li 6.941	4 Be 9.01218											5 B 10.81	6 C 12.011	7 N 14.0067	8 O 15.9994	9 F 18.9984	10 Ne 20.179
3	11 Na 22.9898	12 Mg 24.305											13 Al 26.9815	14 Si 28.0855	15 P 30.9738	16 S 32.066(6)	17 Cl 35.453	18 Ar 39.948
4	19 K 39.0983	20 Ca 40.08	21 Sc 44.9559	22 Ti 47.88	23 V 50.9415	24 Cr 51.996	25 Mn 54.9380	26 Fe 55.847	27 Co 58.9332	28 Ni 58.69	29 Cu 63.546	30 Zn 65.39	31 Ga 69.72	32 Ge 72.59	33 As 74.9216	34 Se 78.96	35 Br 79.904	36 Kr 83.80
5	37 Rb 85.4678	38 Sr 87.62	39 Y 88.9059	40 Zr 91.224	41 Nb 92.9064	42 Mo 95.94	43 Tc (98)	44 Ru 101.07	45 Rh 102.906	46 Pd 106.42	47 Ag 107.868	48 Cd 112.41	49 In 114.82	50 Sn 118.71	51 Sb 121.75	52 Te 127.60	53 I 126.905	54 Xe 131.29
6	55 Cs 132.905	56 Ba 137.33	★	72 Hf 178.49	73 Ta 180.948	74 W 183.85	75 Re 186.207	76 Os 190.2	77 Ir 192.22	78 Pt 195.08	79 Au 196.967	80 Hg 200.59	81 Tl 204.383	82 Pb 207.2	83 Bi 208.980	84 Po (209)	85 At (210)	86 Rn (222)
7	87 Fr (223)	88 Ra 226.025	▲	104 Und (261)	105 Unp (262)	106 Unh (263)	107 Uns (264)	108 Uno (265)	109 Une (266)	110 Uun (267)	111 Uuu (272)	112 Uub (272)	113 Uut	114 Uuq	115 Uup	116 Uuh	117 Uus	118 Uuo

★ Lanthanide Series (RARE EARTH)

57 La 138.906	58 Ce 140.12	59 Pr 140.908	60 Nd 144.24	61 Pm (145)	62 Sm 150.36	63 Eu 151.96	64 Gd 157.25	65 Tb 158.925	66 Dy 162.50	67 Ho 164.930	68 Er 167.26	69 Tm 168.934	70 Yb 173.04	71 Lu 174.967

▲ Actinide Series (RARE EARTH)

89 Ac 227.028	90 Th 232.038	91 Pa 231.036	92 U 238.029	93 Np 237.048	94 Pu (244)	95 Am (243)	96 Cm (247)	97 Bk (247)	98 Cf (251)	99 Es (252)	100 Fm (257)	101 Md (258)	102 No (259)	103 Lr (260)

is the most dense). However, for each Period in which they are found, they are the best oxidizing agents within that Period.

When the halogens are in a gaseous state, they occur as **diatomic** molecules, e.g., Cl_2. But only two of the halogens are gases at room temperature, i.e., fluorine (Fe_2) and chlorine (Cl_2). Bromine is a liquid and iodine is a solid at room temperature. Astatine is the only halogen that is radioactive and is not very important as a representative of the halogens.

FLUORINE (Halogens)
SYMBOL: F   ATOMIC NUMBER: 9   PERIOD: 2

COMMON VALENCE: 1 **ATOMIC WEIGHT:** 18.9984 **NATURAL STATE:** Gas **COMMON ISOTOPES:** Only the isotope with the atomic weight of 19 is stable. Many other unstable radioactive isotopes of fluorine have been artificially produced. **PROPERTIES:** Fluorine, as a **diatomic** gas molecule (F_2), is pale yellow. Fluorine is the most **electronegative** nonmetallic element known (wants to gain electrons) and is, therefore, the strongest **oxidizing agent** known. Since the fluorine atom has only 9 electrons which are close to the nucleus, the positive nucleus has a strong tendency to gain electrons to complete its outer shell. As a gas, its density (specific gravity) is 1.695, and as a liquid, its density is 1.108. Freezing point: –219.61°C; boiling point: –188°C.

CHARACTERISTICS

Fluorine has the lowest atomic weight of all the halogens in Group 17.

Fluorine reacts with most elements except helium, neon, and argon.

It reacts violently with hydrogen-containing compounds, including water and ammonia.

It reacts with metals, such as aluminum, zinc, and magnesium, sometimes bursting into flames. It also reacts with all organic compounds. The results are

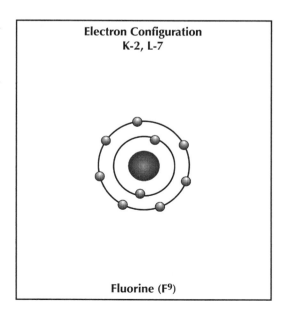

Electron Configuration
K-2, L-7

Fluorine (F^9)

complex fluoride compounds, such as fluorocarbon molecules. Fluorine as a diatomic gas and many fluoride compounds are extremely poisonous and irritating to the skin and lungs. They are also very corrosive.

ABUNDANCE AND SOURCE

Fluorine is the thirteenth most abundant element on the Earth. It makes up about 0.06% of the Earth's crust. Fluorine is widely distributed in many types of rocks and minerals, but never in its pure form since it is too reactive to exist as a free element. Fluorine is as plentiful as nitrogen, chlorine, and copper, but less plentiful than aluminum and iron.

The most abundant fluorine mineral is fluorite—calcium fluoride (CaF_2)—which is often found with other minerals, such as quartz, barite, calcite, sphalerite, and galena. It is mined in Cumberland, England, and in Illinois in the United States.

Other minerals from which fluorine is recovered are fluorapatite, cryolite, and fluorspar, which are found in many countries but mainly in Mexico and Africa.

HISTORY

Fluorine was known for at least fifty years before it was isolated and correctly identified. It was suspected as an element related to chlorine, iodine, and bromine, but it was so reactive that it could not be studied with the equipment available at that time. Before it was identified it was given the name *fluorine* for the Latin word *fluere*, which means "flow."

Even though Carl Wilhelm Scheele (1742–1786) first "discovered" fluorine in 1771, he was not given credit because it was not yet isolated and correctly identified.

In 1886 Ferdinand-Frederic Henri Moissan (1852–1907) used platinum laboratory equipment, which did not react readily with fluorine, to isolate the element and, by passing a current through it, produced fluorine gas (F_2), which could then be accurately identified. Moissan was credited with the discovery of fluorine, and in 1906 he received a Nobel Prize for his work, partly because of the new and unique way in which he produced and identified the element.

COMMON USES

Fluorine compounds are used to reduce the **viscosity** of molten metal and the **slag** by-product so they will flow more easily. Fluorine is also used in the glass and ceramics industries.

One of the most important uses is as various inert fluorocarbon and **chlorofluorocarbon** compounds (CFCs) used as gas propellants in spray cans, e.g., hair spray, deodorant, and paint. They are also used as the coolant in air conditioners/refrigerators. These gases have been banned for use as propellant gas and for use as refrigeration gases (Freon), due to the possible damage to the ozone layer in the upper atmosphere.

Fluorine compounds are also used as additives to drinking water, toothpaste, and dental treatments to reduce tooth decay.

Other uses include nonsticking surfaces (fluoropolymers, called Teflon) for cookware, ironing board covers, razor blades, and so forth.

EXAMPLES OF COMMON COMPOUNDS

Fluorine nitrate (FNO_3)—A strong oxidizing gas or liquid. In the liquid state it explodes by shock or friction. Used as an oxidizer for rocket propellant fuels.

Fluoroacetic acid (CH_2FCOOH)—Very poisonous. Used to kill rats and mice.

Chlorofluorocarbons (CFCs)—There are many forms of fluorocarbons, including the CFCs that are used as refrigerants (Freon) and as the pressure in spray cans. Their use is of concern as being potentially harmful to the ozone layer. In 1987 an international agreement was signed by about ninety nations to reduce the use of CFCs by 50% by the year 2000. This did not seem adequate, so in 1990 a new treaty called for the elimination of the use of all CFCs by all industrial countries.

Today, most air conditioners and refrigerators still use Freon (dichlorodifluoromethane), but manufacturers are slowly changing to systems that do not use CFCs. This creates an economic hardship for persons still using cooling systems that require Freon, because its cost has increased drastically. Over eighty million cars built before 1994 still use CFCs in their air conditioner systems. China, India, Russia, and Mexico still make CFCs. Much of it is illegally smuggled into the United States.

Sodium fluoride (NaF)—1 ppm of NaF is added to drinking water to help reduce dental cavities. Also used as an insecticide, fungicide, and rodenticide, as well as for adhesives, disinfectants, toothpaste, and dental prophylaxis.

HAZARDS

Many of the fluorine compounds (CFCs) are inert and not toxic to humans. But many other types of compounds, particularly the salts and acids of fluorine, are very toxic when either inhaled or ingested. They are also strong irritants to the skin.

There is also the danger of fire and explosion when fluorine combines with several elements and organic compounds.

Poisonous fluoride salts are not toxic to the human body at the very low concentration levels used in drinking water and toothpaste to prevent dental cavities.

❖ ❖ ❖

CHLORINE (Halogens)
SYMBOL: Cl   **ATOMIC NUMBER:** 17   **PERIOD:** 3

COMMON VALENCE: 1, 3, 4, 5, and 7 **ATOMIC WEIGHT:** 35.453
NATURAL STATE: Gas   **COMMON ISOTOPES:** There are 2 stable isotopes of
chlorine: chlorine-35 and chlorine-37. Several artificial radioactive isotopes
have been made. The most important is chlorine-36 with a half-life of about
400,000 years. It is produced by the irradiation of potassium chloride (KCl).
PROPERTIES: Chlorine is a greenish-yellow **diatomic** gas (Cl_2) that is
noncombustible, but readily supports combustion, similar to oxygen. It is
extremely **electronegative** and a strong **oxidizing agent**, but not as strong as
fluorine, although more reactive than the other halogens. As a gas, its specific
gravity (density) is 3.21. As a liquid, it is a clear amber color with a density of
1.56. Its freezing point is –101°C.

CHARACTERISTICS

Chlorine is located between fluorine and bromine in Group 17 of the Periodic Table.

Chlorine gas (Cl_2) has a very pungent odor that is extremely irritating and suffocating when inhaled. In high concentrations it is a deadly poison. Although chlorine is not as reactive as fluorine, it will combine with most other elements.

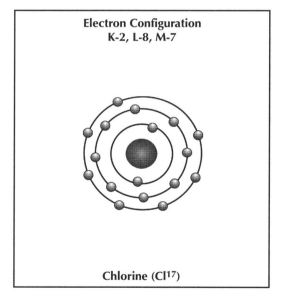

**Electron Configuration
K-2, L-8, M-7**

Chlorine (Cl[17])

ABUNDANCE
AND SOURCE

Chlorine is the twentieth most abundant element on the Earth. It is not found as a free element except as a gas escaping from very hot active volcanos. It does form many compounds.

It has been known as rock salt (halite) for thousands of years. It is also found in sylvite and carnallite and as a chloride ion in seawater.

Chlorine is one of the most commonly produced and used chemicals in the United States. It ranks ninth highest in volume. It is manufactured by

electrolysis of sodium chloride or hydrochloric acid by passing an electric current through solutions of these chlorine compounds.

It is also produced by the **oxidation** of hydrogen chloride (HCl) with nitrogen oxide (NO_x).

HISTORY

In 1774 Carl Wilhelm Scheele (1742–1786) isolated chlorine, a greenish-yellow gas, whose color was unexpected because the gases he was working with were usually colorless. He named it *chlorine* after the Greek word *khloritis,* which means "green stone." Scheele isolated several other elements and compounds, including oxygen, hydrogen fluoride, hydrogen sulfide, and hydrogen cyanide.

Some references give Scheele credit for the discovery of chlorine. Others say that he did not recognize the gas as a new element (he thought it was something combined with oxygen) and thus do not give him credit. Some references list Sir Humphry Davy, who correctly identified chlorine in 1810, as the discoverer.

COMMON USES

There are many uses for chlorine. Some common ones are the manufacture of carbon tetrachloride, many chlorinated hydrocarbons, hydrogen chloride, and hydrochloric acid.

Chlorine is used to make some plastics such as neoprene and polyvinyl chloride (vinyl).

It is used for water purification, bleaching, and the processing of meat, fish, vegetables, and fruit.

Some of the compounds are used to make **insecticides**, fireworks, explosives, textile processing and dyes, paint **pigments**, **pharmaceuticals**, chloroform, and **chlorofluorocarbons** (CFCs) and **chlorohydrocarbons** (CHCs).

Chlorine was also used as a poisonous gas during World War I.

EXAMPLES OF COMMON COMPOUNDS

Carbon tetrachloride (CCl_4)—Years ago it was mixed with ether and sold as Carbona, a dry-cleaning fluid for clothes. It is no longer permissible to sell or buy CCl_4 for household use. It is classed as a **carcinogen** by the U.S. government and is toxic if ingested, inhaled, or absorbed by the skin. Carbon tetrachloride is used to manufacture CFHCs, to fumigate grains to kill insects, and in the production of **semiconductors**.

Ethyl chloride (C_2H_5Cl)—Used to manufacture tetraethyl lead, as a refrigerant, and as an anesthetic.

Chlorine dioxide (ClO_2)—A reddish-yellow gas used as a bleach for foods and to purify swimming pools.

Phosgene ($COCl_2$)—A very poisonous gas used in wartime. When not concentrated, it smells like newly cut hay or grass.

Sodium hypochlorite (NaOCl)—When concentrated, it is a strong oxidizer and may explode. Used for swimming pools, and when diluted to 5.25%, as a laundry bleach (Clorox).

Chlorinated hydrocarbons—One example, DDT, is an insecticide. It was used extensively in World War II to delouse personnel and prevent the spread of plague and other insect-borne diseases. Today, its use is restricted because of its toxicity and its very long life. It is difficult to eliminate in nature.

Chloroform ($CHCl_3$)—Toxic and **carcinogenic** if ingested or inhaled over a long period of time. Formerly used as an **anesthetic** during surgical procedures. Today, it is banned for use in cosmetics and items such as toothpaste and cough syrup.

HAZARDS

A series of chlorofluorohydrocarbons that are used as refrigerants are being phased out of manufacture and use, because of their possible effects on the ozone layer of the atmosphere (see Oxygen).

From time to time a railroad tank car involved in an accident will leak liquid or gaseous chlorine which, when escaping into the air, forms chlorine toxins. This is dangerous, both as a fire hazard and for human health. When water is used to flush away the escaping chlorine, it may end up as hydrochloric acid, which can be hazardous to the water supply and to fish.

Concentrated chlorine gas and many chlorine compounds will **oxidize** powdered metals, hydrogen, and many organic materials, and release enough heat to start fires or explode.

Chlorine is constantly evaporating from the oceans and drifting into the atmosphere where it causes a natural depletion of the ozone.

Never mix or use together chlorine cleaners, such as Clorox, with other cleaners containing ammonia. It's a deadly mixture.

BROMINE (Halogens)
SYMBOL: Br   ATOMIC NUMBER: 35   PERIOD: 4

COMMON VALENCE: 1, 3, 5, and 7   **ATOMIC WEIGHT:** 79.904   **NATURAL STATE:** Liquid   **COMMON ISOTOPES:** Bromine has 2 stable isotopes, bromine-79 and bromine-81, which occur in about equal proportions in nature. There are several known radioisotopes. **PROPERTIES:** Bromine is a thick, dark-red liquid with a high density. It is the only nonmetallic element that is a liquid at normal room temperature. Density as a solid: 4.05, as a liquid: 3.123; boiling point: 58.8°C; freezing point: –7.27°C.

CHARACTERISTICS

Bromine is a very reactive nonmetallic element, located between chlorine and iodine in the Periodic Table.

Bromine gas fumes are very irritating and toxic.

Bromine is soluble in most organic **solvents** and only slightly soluble in H_2O. Liquid bromine will attack most metals, even platinum.

ABUNDANCE
AND SOURCE

Bromine is the sixty-third most abundant element. Although it is not found uncombined in na-

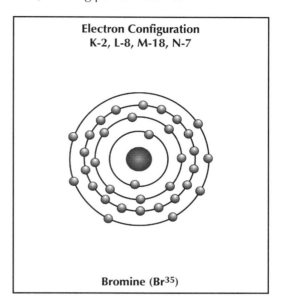

Electron Configuration
K-2, L-8, M-18, N-7

Bromine (Br^{35})

ture, it is widely distributed over the Earth in low concentrations. It is found in seawater at a 65 ppm concentration. This concentration is too low to be extracted directly, so the salt water must be concentrated, along with chlorine and other salts, by **solar evaporation** and/or **distillation**.

Most of the bromine is recovered from underground brines, from the Great Salt Lake of Utah in the United States, and from Stassfurt, Germany. It is commercially produced, along with potash, from evaporation of the high-salt-content water of the Dead Sea, which is 1,290 feet below sea level and is located on the borders of the countries Israel and Jordan.

HISTORY

Bernard Courtois (1777–1838), a manufacturer of gunpowder and explosives, experimented with seaweed to extract some chemicals he might have been able to use as explosives. He discovered iodine, identified by the violet color of its gas, but he still had a brownish-red residue left over, which he assumed was a different element. Since his new discovery had such a strong odor, he named it *bromine* from the Greek word *bromos*, which means "stench or smell."

Some sources say that, in 1825, Antonine J. Balard (1802–1876) discovered the liquid element bromine but did not publish his findings until later. Several chemists were working with the halogens during the same time period, including Carl Lowig (1803–1890) and Justus von Liebig (1803–1873).

COMMON USES

Bromine has many uses related to its unique properties and the way it reacts with organic materials.

It is used to make an additive for **antiknock** gasoline, as a bleaching agent, to purify water, as a fire retardant, and to make dyes and **pharmaceutical** products. It is also used in photography and as a flame retardant.

Besides the inorganic uses, bromine has many unique applications with organic compounds. It is used as a **reagent** to study the organic reactions of many other compounds. It is used as a disinfectant, a fumigant, a sedative, and to purify water.

EXAMPLES OF COMMON COMPOUNDS

Bromine azide (BrN_3)—A strong **oxidizing agent** that will explode when shocked or heated. Used to make detonators for dynamite and other explosives.

Bromine chloride (BrCl)—An irritating red liquid used to treat industrial and sewage wastes.

Bromine pentafluoride (BeF_5)—Very **corrosive** to the skin and explodes when in contact with water. Used as an oxidizer in rocket fuel.

Bromophosgene (carbon oxybromide) ($COBr_2$)—Toxic if inhaled or ingested. Used as a military poisonous gas.

HAZARDS

Bromine liquid and many bromine compounds in liquid form are very difficult to remove from the skin. They will cause serious burns that take a long time to heal.

Almost all bromine compounds are toxic if inhaled or ingested. Many are extremely explosive. When working with bromine, you must be very careful not to breathe the vapors or get any on your skin.

IODINE (Halogens)
SYMBOL: I   **ATOMIC NUMBER:** 53   **PERIOD:** 5

COMMON VALENCE: 1, 3, 5, and 7 **ATOMIC WEIGHT:** 126.905
NATURAL STATE: Solid **COMMON ISOTOPES:** There are no stable isotopes of iodine, but the elemental form is iodine-127 with 53 protons and 74

neutrons. There are at least 22 radioactive artificial isotopes of iodine. Only iodine-131 with a **half-life** of 8 days is of importance, particularly in medicine. **PROPERTIES:** Iodine exists as a nonmetallic **diatomic** molecule (I_2). Iodine is the heaviest of the naturally occurring halogens. It is a black solid with a luster and has an easily identifiable odor when used as an **antiseptic**. Density: 4.95; melting point: 113.5°C; boiling point: 184°C.

CHARACTERISTICS

Iodine is the least reactive of the halogens.

Like other halogens, iodine can form both ionic and covalent bonds.

Iodine **sublimates** (turns from a solid directly to a gas without going through a liquid state) to form a violet gas for which it is named.

ABUNDANCE AND SOURCE

Iodine is the sixty-fourth most abundant element. It occurs widely over the Earth, but never in the elemental form and never in high concentrations.

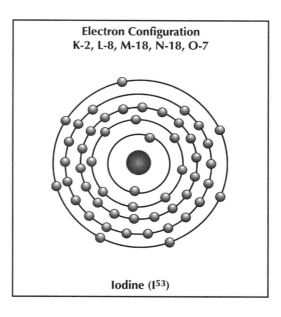

Electron Configuration
K-2, L-8, M-18, N-18, O-7

Iodine (I^{53})

It occurs in seawater where some species of seaweed and kelp accumulate the element in their cells. It is recovered from brine wells found in Chile, Indonesia, Japan, and Michigan and Oklahoma in the United States. It is also extracted from seaweed and kelp.

HISTORY

Bernard Courtois (1777–1838) experimented with seaweed to produce potassium carbonate, used in the making of explosives. When he heated the mixture, a violet vapor appeared that, when condensed, formed black crystals. Sir Humphry Davy (1778–1829) confirmed this discovery of a new element and named it *iodine* after the Greek word, *iodes*, which means "violet," but Courtois was given credit for the discovery of iodine.

COMMON USES

One of the most important uses of iodine is in the treatment of hypothyroidism, a condition where the thyroid gland does not receive adequate iodine. Iodine deficiency may lead to the formation of a goiter, where the gland that surrounds the windpipe in the neck becomes enlarged. There are other causes of goiter, including cancer of the thyroid gland.

A deficiency of iodine can also cause cretinism (infant hypothyroidism) in newborn babies, which can result in mental retardation unless the subject takes thyroid hormones for a lifetime. Some foods grown in iodine-deficient soils do not contain adequate iodine for our diets. This is why, many years ago, iodine was added to salt for people who live in regions with iodine-poor soils. The area around the Great Lakes in the United States is one region with soil deficient in iodine. A healthy diet requires 90 to 150 micrograms of iodine a day, which, in addition to iodized salt, can be obtained from eating a balanced diet, including seafood.

Iodine-131 is an artificial **radioisotope** of iodine used as a **tracer** in biomedical research and as a treatment for thyroid diseases.

In industry, iodine is used for dyes, antiseptics, germicides, X-ray contrast medium, food and feed additives, **pharmaceuticals**, medicinal soaps, photography film emulsions, and as laboratory catalysts to either speed up or slow down **chemical reactions**.

EXAMPLES OF COMMON COMPOUNDS

Iodine-131—Is produced artificially in a nuclear pile by **radiation** of tellurium or is recovered as an end **fission** product in a **nuclear reactor**. Its 8-day half-life makes it a useful research, diagnostic, and treatment radioisotope.

Iodine tincture—A well-known antiseptic form of iodine. A tincture is a solution of iodine that has been dissolved in alcohol. As an antiseptic it consists of about 1/2 alcohol, and about 1/2 water, with only a small amount of iodine.

Iodine monochloride (ICl)—A black to reddish crystal that is toxic if inhaled or ingested. Used to produce organic compounds.

HAZARDS

Elemental iodine and most of its compounds, as gases, liquids, or solids, are toxic if inhaled or ingested. In high concentrations most iodine compounds also cause severe irritation to the skin.

Like many antiseptics and germicides, iodine is a poison. However, in diluted forms, iodine is safe and very beneficial to humans.

Iodine, although a poison in high concentrations, is required as a trace element in our diets to prevent thyroid problems and mental retardation.

❖   ❖   ❖

ASTATINE (Halogens)
SYMBOL: At   **ATOMIC NUMBER:** 85   **PERIOD:** 6

COMMON VALENCE: 1, 3, 5, and 7 **ATOMIC WEIGHT:** 211 **NATURAL STATE:** Solid **COMMON ISOTOPES:** There are no stable isotopes. About 25 radioactive isotopes have been artificially prepared. They range from astatine-196 to astatine-223. Most have very short **half-lives**. Astatine-210 with an 8.3-hour half-life and astatine-211 with a 7.2-hour half-life are used as radioactive tracers. **PROPERTIES:** Astatine is the halogen that most resembles a metal. Most properties of astatine, such as density, melting point, and boiling point, are difficult to determine because of the short existence of the isotopes. Estimated melting point: 320°C; estimated boiling point: 336.9°C.

CHARACTERISTICS

Astatine is the heaviest and most dense of the non-metallic elements in Group 17 (VIIA). Its physical and chemical properties are thought to be similar to those of iodine.

ABUNDANCE AND SOURCE

Astatine is found only in trace amounts. It is the rarest element on Earth. There is only about 1 ounce of astatine in existence on the entire Earth in its natural but unstable state. It is found in trace amounts in uranium minerals and ores.

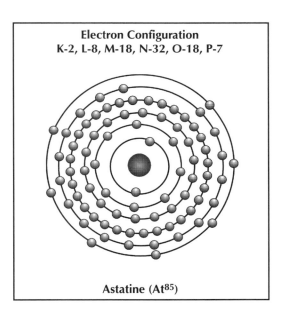

Electron Configuration
K-2, L-8, M-18, N-32, O-18, P-7

Astatine (At⁸⁵)

Since its half-life is very short, there would be none left on Earth if it were not replenished. Astatine is a by-product of the radioactive decay process of uranium, which keeps producing a new trace supply that soon decays.

HISTORY

In 1940, just before World War II, three scientists, Dale Raymond Corson (1914–1995), K. R. Mackenzie (1912–1995) and Emilio Gino Segre (1905–1989), produced a new element with 85 protons by bombarding bismuth with **alpha particles** (helium nuclei). Their discovery was not confirmed until later because of the interruption of the war. Astatine filled in the last gap of the Periodic Table, except for Element 61, which had not yet been discovered.

Because of its natural radioactivity, it received its name, *astatine*, from the Greek word *astatos*, which means "unstable," since it is the only halogen without a stable isotope.

Again, there is some question as to who discovered astatine. Some authorities state that Fred Allison and E. J. Murphy discovered astatine in 1931, but most give Corson, Mackenzie, and Segre the credit.

COMMON USES

In water solution astatine resembles iodine in some of its chemical properties. Both are powerful **oxidizing agents**.

It has limited use in medicine as a radioactive source. It concentrates in the thyroid gland just like iodine, which makes it a useful **radioisotope tracer**.

Astatine has very few other practical uses.

EXAMPLES OF COMMON COMPOUNDS

Astatine does not have many stable or useful compounds. Like a halogen, it will form halogen salts with some other elements.

Astatine-211, which has a half-life of just over 7 hours, is used as a radioactive tracer for thyroid diseases.

HAZARDS

The major hazard is from the radiation of astatine's isotopes. However, since these isotopes have very short half-lives, they do not pose a great long-term danger. Even so, astatine is considered a dangerous element that is a radioactive poison and carcinogen. It has been demonstrated that astatine causes cancer in laboratory animals.

Group 18 Elements
THE NOBLE GASES

The noble gases are found in Group 18 (VIIIA), which, according to some older versions of the Periodic Table, is called Group 0. Having a full outer valence shell, they represent the end of each period of the Table. They are neon, argon, krypton, xenon, and radon. The noble gases are colorless, tasteless, and odorless. They glow in glass tubes when electric currents are passed through them. Since each inert gas glows with its own distinctive color, the color of the glowing light is dependent on the mixture of the gases.

The noble gases are sometimes referred to as "rare gases," which is not exactly accurate since argon makes up almost 1% of our atmosphere. Sometimes they are called "inert gases," which is not exactly correct either, since, under very specific conditions of temperature and pressure, some of them will combine with oxygen or the more reactive halogens to form compounds.

The reason for the lack of chemical activity of the noble gases is that their outer valence shells have a full complement of 8 electrons. They do not need to give up, receive, or share electrons with other elements. Their electron configuration is the most stable of all the elements.

Helium (He^2) is not really one of the noble gases, but because it is inactive and has a completed first shell (K = 2 electrons), it is placed at the top of Group 18 (VIIIA). Radon, which is the largest member of the noble gases, is the only radioactive noble gas. Its outer electron shell is furthest from the nucleus.

HELIUM (Noble Gases)
SYMBOL: He   ATOMIC NUMBER: 2   PERIOD: 1

COMMON VALENCE: 0   ATOMIC WEIGHT: 4.0026   NATURAL STATE: Gas
COMMON ISOTOPES: Helium-3 (3 neutrons), Helium-4 (4 neutrons), and helium-5 are extremely rare isotopes that exist in very minute quantities. Helium-6 and helium-8 are **radioactive isotopes** that decay by emitting **beta particles** to form lithium-6 and lithium-8. **PROPERTIES:** Helium is a colorless, odorless, tasteless, inert gas. It is noncombustible, and it is the least soluble of any gas in water and alcohol. As a gas, it diffuses well in solids.

Helium has the lowest freezing point of any substance, i.e., –272.2°C. At near absolute zero it exhibits superfluidity with a high thermal conductivity. **Absolute zero** is –273.13°C or –459.4°F, where all molecular/thermal motion ceases. Its density (specific gravity) as a gas is 0.17847; density as a liquid is 0.1249; freezing point: –272.2°C; boiling point: –268.94°C.

CHARACTERISTICS

The helium atom consists of a nucleus containing 2 protons and 2 neutrons, with 2 electrons in the first (K) shell. Since its outer shell is complete, it is considered inactive and is the most inert of the noble elements.

The nuclei of helium atoms are called **alpha particles**, each of which has a charge of +2 and an atomic weight of 4.

Liquid helium exhibits some unusual characteristics when supercooled.

First, it is the only element that will not turn to a solid by just using pressure. Heat must be removed as the pressure is increased, but helium will freeze at –272.2°C, which is the lowest temperature scientists have ever achieved.

Second, it is an excellent conductor of heat. As a supercold liquid, it will move toward heat—even flow up the sides and over the top of a container.

SOURCE AND ABUNDANCE

Helium is the seventy-first most abundant element on Earth, but it is the second most abundant element in the universe, after hydrogen. Most likely helium was the first element to be formed after hydrogen during the big bang. The theory

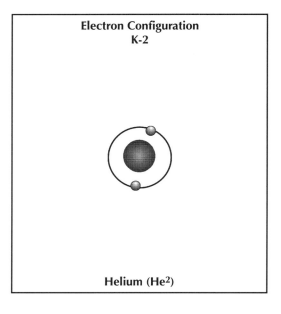

Electron Configuration
K-2

Helium (He^2)

is that hydrogen atoms combined under heat and pressure to form helium atoms. The Earth's current helium originally came from the natural decay of radioactive elements deep in the Earth. Much of it seeps up to the surface and escapes into the atmosphere. And, like hydrogen, it is a very light gas that sooner or later escapes from the Earth's gravity.

Helium can be obtained from the atmosphere by lowering the temperature of air until it liquefies. All the other gases in air will turn to a liquid except helium, because it has the lowest boiling point. Since helium, at this stage of cooling, will be the only vapor left, it can be removed as a pure gas.

It is commercially more profitable to produce helium by separating it from a mix of natural underground gases where its concentration is greater than in the atmosphere. It is produced in Oklahoma, Kansas, New Mexico, Arizona, and Canada by liquefying natural gas and "boiling off" the other gases. Helium is then purified to 99.995%. Amarillo, Texas, is one of the major centers for the production of helium in the United States. Most of the world's supply of helium comes from the United States.

HISTORY

Helium was identified, independently, by Pierre-Jules-Cesar Janssen (1824–1907) and Joseph Norman Lockyer (1836–1920) as part of the sun's atmosphere by studying sunlight during a solar eclipse in India through a **prism**. It was also discovered on Earth by William Ramsay (1852–1916), who was given credit for the discovery, as was Per Theodor Cleve

(1840–1905), along with Niles Langlet, who discovered it independently. It was not only the first inert gas discovered, it was the first element discovered in the sun before it was identified on the Earth. Helium was one of the new elements predicted by the Periodic Table.

Because it was observed first in sunlight, Edward Frankland (1825–1899) named the element *helium*, from the Greek word *helios*, which means "sun."

COMMON USES

Helium has many uses:

As an inert gas, it is used as the atmosphere in which to "grow" silicone crystals (computer chips).

As a lifting gas, it is used to inflate weather balloons and lighter-than-air ships similar to the ones seen taking TV pictures above football games. Helium is used for all lighter-than-air ships, even though it has less lifting power than hydrogen. Being noncombustible, it is safer than hydrogen.

In arc welding, it is used as an inert gas shield which releases great heat for very long and heavy welds. Helium prevents oxidation of the metal being welded, thus preventing burning and corrosion of the metal. This is one of the major uses of helium.

Helium is used for low-temperature research ($-272.2°C$ or $-434°F$). It is becoming important as a coolant for **superconducting** electrical systems which, when cooled, offer little resistance to the electrons passing through a conductor (wire or magnet). When the electrons are "stripped" from the helium atom, a positive He^{++} ion results. The positive helium ions (nuclei) occur in both natural and man-made radioactive emissions and are referred to as **alpha particles**. Helium ions (alpha particles) are used in high-energy physics to study the nature of matter.

In gas discharge **lasers**, helium transfers the energy to the laser gas such as carbon dioxide or another inert gas.

As an inert gas with heat-transfer capability, helium is used in gas-cooled nuclear power reactors, which operate at a higher efficiency than liquid-cooled nuclear reactors.

Helium is mixed with oxygen in scuba-diving and deep-sea diving air tanks because it is less soluble in divers' blood than nitrogen. Divers have a greater chance of experiencing "nitrogen narcosis" and becoming disoriented, and also of getting "the bends" when using compressed air (nitrogen-oxygen mixture), a condition where the nitrogen forms bubbles in the blood as divers ascend. This condition is not only painful but life-threatening, particularly if divers become so disoriented that they don't know where they are, or if they come to the surface too fast. The chances of deep-sea divers getting the bends and becoming disoriented are lessened when breathing a helium-oxygen mixture because helium is less soluble in the blood than nitrogen.

EXAMPLES OF COMMON COMPOUNDS

As an inert element (full outer K shell of 2 electrons), it does not easily form compounds. Its valence is 0. Thus, it does not react with other elements. It is classified as the most inert of the noble elements in Group 18 [VIIIA].

Helium is also the result of **fusion** reactions where the nuclei of heavy hydrogen are "fused" to form atoms of helium. The result is the release of great amounts of energy. Fusion is the physical or nuclear reaction (not chemical reaction) that takes place in the sun and in thermonuclear weapons, e.g., the hydrogen bomb.

HAZARDS

Helium is not toxic, but it is an **asphyxiant** gas—if you breathe it, you will not be killed by the gas, but rather by oxygen deprivation.

A possible hazard is when He++ nuclei, as alpha particles, are accelerated to high speeds and bombard a target. Alpha particles can be stopped by several inches of air or a piece of cardboard. As high-energy, charged particles generated from man-made or natural radioactivity, alpha particles can cause tissue damage, but they are not as damaging to our bodies as are the very short wavelength **gamma rays**, which can be stopped by lead shielding.

NEON (Noble Gases)
SYMBOL: Ne   ATOMIC NUMBER: 10   PERIOD: 2

COMMON VALENCE: 0 **ATOMIC WEIGHT:** 20.18 **NATURAL STATE:** Gas
COMMON ISOTOPES: There are 3 stable isotopes. The most abundant is neon-20. The others are neon-21 and neon-22. **PROPERTIES:** Neon is a colorless, odorless, tasteless gas that does not form compounds. It will glow bright red when electricity is passed through it in an enclosed glass tube. Density: 0.6964 (air = 1); freezing point: –248.6°C; boiling point: –246.1.

CHARACTERISTICS

Neon is the second element in Group 18 (VIIIA) of the Periodic Table.

Neon will turn to a liquid at –245.92°C. It is noncombustible and lighter than air, but not as light as helium.

ABUNDANCE AND SOURCE

Neon is the eighty-second most abundant element on Earth. It is found in very small amounts in the atmosphere.

Neon is believed to be produced by **radioactive** decay deep in the Earth. As it rises to the surface, it escapes into the atmosphere and is soon dissipated. Some neon is found mixed with natural gas and several minerals.

Neon is recovered from liquid air by **fractional distillation**. As the liquid turns back into a gas, the neon is separated from the other gases in the air.

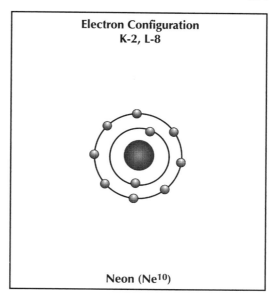

**Electron Configuration
K-2, L-8**

Neon (Ne^{10})

HISTORY

William Ramsay (1852–1916), who received the Nobel Prize in 1904 for his discovery of argon, also identified neon, which he named after the Greek word *neos*, meaning "new." Morris William Travers (1872–1961) shares the credit for the discovery of neon because he identified neon at the same time as Ramsay. They used fractional distillation of liquid air, which allows different gases to return from the liquid state to the gaseous state at different temperatures. Thus, the gases can be collected separately as they "boil off" at their own temperature. These scientists made their identifications of the noble gases by use of a **spectrometer**, which identifies each element by its unique lines in the spectrum of light as it is excited, heated, or burned.

COMMON USES

Because of neon's ability to ionize in an electrical current, it can be used to make **luminescent** electric tubes and specialty high-voltage indicators. It produces a reddish/orange glow when ionized, which we know as "neon lights" used for commercial advertising.

EXAMPLES OF COMMON COMPOUNDS

None.

HAZARDS

Neon is nontoxic. As an **asphyxiant** gas it can smother you by taking the place of oxygen in your lungs.

ARGON (Noble Gases)

SYMBOL: Ar   **ATOMIC NUMBER:** 18   **PERIOD:** 3

COMMON VALENCE: 0   **ATOMIC WEIGHT:** 39.948   **NATURAL STATE:** Gas
COMMON ISOTOPES: There are 3 stable isotopes of argon. Most of the argon gas found in the atmosphere is argon-40; the other stable isotopes are argon-36 and argon-38. **PROPERTIES:** Argon is a colorless, odorless, tasteless inert gas that does not combine with other elements. It can be liquefied. Density: 1.38 (air = 1); freezing point: −189.4°C; boiling point: −185.9°C.

CHARACTERISTICS

Argon is not known to combine with any other element.

It is only slightly soluble in water.

ABUNDANCE AND SOURCE

Argon is the fifty-sixth most abundant element on Earth. It is the most abundant of the noble gases found in the atmosphere. In fact, the only source of argon is the atmosphere, where it is found at just under 1% of air by volume.

There are several methods of producing argon. The most common is by **fractional distillation** of liquid air. Argon boils off from the liquid air at its own unique temperature. Then the argon is collected at a temperature that is between the boiling and condensation temperatures of nitrogen and oxygen gases.

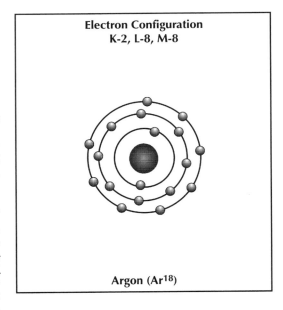

Electron Configuration
K-2, L-8, M-8

Argon (Ar¹⁸)

HISTORY

Lord Rayleigh (1842–1919) isolated a gas that would not combine with oxygen, but he could not identify it. With the assistance of William Ramsay (1852–1916), Rayleigh identified the gas and named it *argon*, after the Greek word *argos*, which means "inert." Both were given credit for the discovery of argon. It was identified by the new technology called **spectroscopy**. The **spectroscope** can identify each element, when heated, by the color and lines it forms on a light spectrum. Each element's spectrum is unique.

COMMON USES

Argon's main uses are as an inert gas, or in a mixture with other inert gases, to fill electric light bulbs, fluorescent tubes, lasers, and so forth. It is also used as a nonoxidizing gas for welding and to decarbonize steel.

EXAMPLES OF COMMON COMPOUNDS

None.

HAZARDS

Argon is nontoxic, but as an **asphyxiant** gas, it can smother you by taking the place of oxygen in your lungs.

KRYPTON (Noble Gases)
SYMBOL: Kr   **ATOMIC NUMBER:** 36   **PERIOD:** 4

COMMON VALENCE: 0 (2)   **ATOMIC WEIGHT:** 83.80   **NATURAL STATE:** Gas   **COMMON ISOTOPES:** Krypton has 6 stable isotopes and several man-made radioactive isotopes, the most important of which is krypton-85 with a half-life of 10.3 years.   **PROPERTIES:** Krypton is a colorless, odorless, tasteless noble gas, but it does combine with fluorine with a valence of 2, at very low liquid temperatures. Density: 2.818 (air = 1); freezing point: –157°C; boiling point: –153°C.

CHARACTERISTICS

Krypton is the fourth element in Group 18 (VIIIA, which is also known as Group 0).

Even though krypton is a noble, inert gas, several compounds have been formed.

ABUNDANCE AND SOURCE

Krypton is the eighty-first most abundant element on Earth. It makes up a very small portion of our air, about 0.000108% by volume. There are traces of krypton in some minerals and meteorites.

Krypton is found outside the Earth in space. (Shades of Superman?)

HISTORY

William Ramsay (1852–1916) had earlier discovered argon and helium. With the assistance of his partner, Morris William Travers (1872–1961), he discovered neon and xenon, as well as krypton, which was named after the Greek word *kryptos*, which means "hidden."

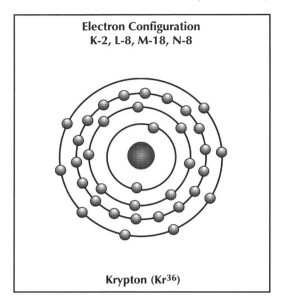

Electron Configuration
K-2, L-8, M-18, N-8

Krypton (Kr³⁶)

COMMON USES

Krypton is used in a mixture with argon to fill **fluorescent** lamps, **incandescent** bulbs, **lasers**, and high-speed photography lamps.

The definition of the metric system's "meter" as a standard length of measurement is based on the spectrum of the krypton-86 isotope.

EXAMPLES OF COMMON COMPOUNDS

Even though krypton is inert, at very low temperatures (–220°C) it will combine with fluorine to make KrF_2 and KrF_4 compounds that decompose at higher temperatures.

HAZARDS

Krypton, being an inert gas, is nontoxic. However, the man-made radio-isotopes of krypton can cause radiation poisoning.

❖   ❖   ❖

XENON (Noble Gases)
SYMBOL: Xe   **ATOMIC NUMBER:** 54   **PERIOD:** 5

COMMON VALENCE: 2, 4, 6, and 8   **ATOMIC WEIGHT:** 131.30   **NATURAL STATE:** Gas   **COMMON ISOTOPES:** There are 9 stable isotopes of xenon. At least 16 radioactive isotopes are known.   **PROPERTIES:** Xenon is a colorless, odorless, tasteless gas that is classed as a noble gas, but it is not completely inert. Xenon is noncombustible. Density: 5.897; boiling point: –108.12°C; will form a liquid at –106.9°C.

CHARACTERISTICS

Xenon, though considered inert, will combine with a few elements, particularly fluorine and oxygen. Xenon exhibits all of the even valence states (2, 3, 4, and 8).

ABUNDANCE AND SOURCE

Xenon is found in trace amounts in the atmosphere. It makes up just 0.086 ppm by volume of air.

The only commercial source of xenon is **fractional distillation** of liquid air. It is also found in some minerals and meteorites.

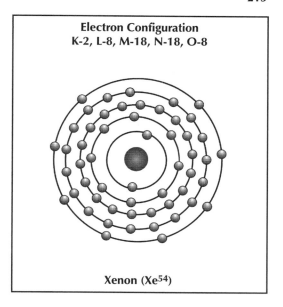

Electron Configuration
K-2, L-8, M-18, N-18, O-8

Xenon (Xe^{54})

HISTORY

Xenon was one of several noble gases discovered by William Ramsay (1852–1916) and his assistant, William Morris Travers (1872–1961). Xenon is one of the six inert noble gases of Group 18 (VIII). They are He, Ne, Ar, Kr, Xe, and Rn.

Their new discovery, the inert gas xenon, was named for the Greek word *xenos*, which means "strange."

COMMON USES

Xenon, mixed with argon, is used in **luminescent** tubes, **lasers**, **tracer** studies, and **anesthesia**. It makes an excellent electronic speed light for photography because of its bright, well-balanced white light and the thousands of times it can be flashed.

EXAMPLES OF COMMON COMPOUNDS

Not much is known about xenon compounds.

Two that are known are xenon combined with fluoride, called xenon hexafluoroplatinate, and xenon oxides, which are stable at room temperatures but not of much use.

HAZARDS

As a noble gas that is mostly inert, xenon is nontoxic and noncombustible. Some of its compounds are toxic, but there is not much chance of coming into contact with them.

❖ ❖ ❖

RADON (Noble Gases)
SYMBOL: Rn   **ATOMIC NUMBER:** 86   **PERIOD:** 6

COMMON VALENCE: 2, 4, and 6 **ATOMIC WEIGHT:** 222 **NATURAL STATE:** Gas **COMMON ISOTOPES:** There are no stable isotopes. There are 3 natural radioactive isotopes, plus 22 radioactive isotopes artificially made by the decay of other radioactive elements in nuclear reactors. All have very short half-lives.

The radioisotope radon-222 has a half-life of 3.8 days and emits high-energy **alpha radiation** or particles (nuclei of helium atoms). **PROPERTIES:** Radon is a colorless, odorless, tasteless gas that is highly radioactive. Density: 9.72; freezing point: –61.8°C, where it becomes a glowing solid. In the liquid state, it is colorless and transparent.

CHARACTERISTICS

Radon's characteristics place it in the group with the other inert gases.

It is the heaviest of the noble gases and is the only one that is radioactive. It emits **alpha particles**.

It can form a limited number of compounds at high temperatures.

ABUNDANCE AND SOURCE

There is only a trace of radon in the atmosphere. It is only made by the natural decay process from radioac-

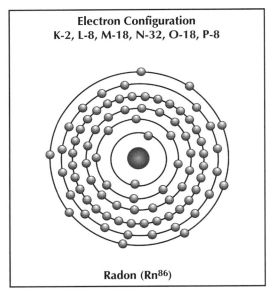

Electron Configuration
K-2, L-8, M-18, N-32, O-18, P-8

Radon (Rn[86])

tive uranium-238 → radium → radon. Radon is constantly being formed in the Earth and escapes to the surface through porous soils and crevices.

Radon gas is recovered by forcing air through radon salt solutions and then collecting the gas.

HISTORY

In 1900 Friedrich Ernst Dorn (1848–1916), while studying radium, found that it gave off a radioactive gas which, when studied in more detail, proved to be the sixth noble gas. Dorn was given credit for the discovery of radon even though it was also identified as "thoron," which is a radioisotope of radon, by several other scientists.

Because of its radioactivity it was named *radon*, from the Latin word *radius*, which means "ray."

COMMON USES

Radon is mostly used for treating cancers. It is also used to trace leaks in gas and liquid pipelines and to measure their rate of flow.

Its escape rate from known sources in cracks of the Earth is used to help predict earthquakes.

EXAMPLES OF COMMON COMPOUNDS

Since radon is inert, but radioactive, there are not many useful compounds.

HAZARDS

Radon is a major radiation hazard.

Homes with cracked concrete slabs or basements may allow radon generated in the Earth to seep through. If the lower levels of homes are not well ventilated, excessive radon may collect, which would expose occupants to excessive radiation. Simple test kits are available for homeowners to check the level of radon in their homes. The seriousness of radon escaping into homes and other buildings is not established. Some authorities attribute the increase in lung cancer in non-cigarette-smoking people to radon gas poisoning.

Several studies found no positive correlation between the rate of cancer and the number of homes that had radon gas exposure for large groups of individuals in several countries. On the other hand, the U.S. Environmental Protection Agency (EPA), which uses the controversial statistical technique of "linear no-threshold extrapolation," insists that any exposure to radon increases the risk of cancer.

9

LANTHANIDE SERIES (RARE-EARTH ELEMENTS)

The lanthanide series is composed of elements with similar physical properties, chemical characteristics and unique structures. They are found in Period 6, starting at Group 3 of the Periodic Table.

The elements in the **lanthanide series** are also called **rare-Earth elements**, not because they are scarce or rare, but because at one time they were thought to be rare because they were very difficult to extract from their ores, difficult to separate from each other, and difficult to identify. Chemical elements that have similar physical and chemical properties tend to occur together in the same ores and minerals.

The lanthanide series can begin either with lanthanum (La^{57}) or with cerium (Ce^{58}) and continue through lutetium (Lu^{71}).

The element yttrium (Y^{39}), located just above lanthanum in Group 3, is sometimes included in this series because its physical properties and chemical characteristics are similar to those of other elements in the series.

We have chosen to start this series with lanthanum, which is located in Group 3 of Period 6. Lanthanum is not always considered as belonging to the series that bears its name, but it closely resembles the other elements in the series. Thus, this is a logical place to start the series.

As mentioned, all of the elements in the lanthanide series possess similar physical properties and chemical characteristics. One of the major properties of these elements is that their **valence** electrons are not in their outer shells. In all fifteen of the lanthanides the outer shell is the sixth or "P" shell,

PERIODIC TABLE OF THE ELEMENTS

GROUPS	1 IA	2 IIA	3 IIIB	4 IVB	5 VB	6 VIB	7 VIIB	8	9 VIII	10	11 IB	12 IIB	13 IIIA	14 IVA	15 VA	16 VIA	17 VIIA	18 VIIIA
PERIODS																		
1	1 H 1.0079																	2 He 4.00260
2	3 Li 6.941	4 Be 9.01218											5 B 10.81	6 C 12.011	7 N 14.0067	8 O 15.9994	9 F 18.9984	10 Ne 20.179
3	11 Na 22.9898	12 Mg 24.305											13 Al 26.9815	14 Si 28.0855	15 P 30.9738	16 S 32.066(6)	17 Cl 35.453	18 Ar 39.948
4	19 K 39.0983	20 Ca 40.08	21 Sc 44.9559	22 Ti 47.88	23 V 50.9415	24 Cr 51.996	25 Mn 54.9380	26 Fe 55.847	27 Co 58.9332	28 Ni 58.69	29 Cu 63.546	30 Zn 65.39	31 Ga 69.72	32 Ge 72.59	33 As 74.9216	34 Se 78.96	35 Br 79.904	36 Kr 83.80
5	37 Rb 85.4678	38 Sr 87.62	39 Y 88.9059	40 Zr 91.224	41 Nb 92.9064	42 Mo 95.94	43 Tc (98)	44 Ru 101.07	45 Rh 102.906	46 Pd 106.42	47 Ag 107.868	48 Cd 112.41	49 In 114.82	50 Sn 118.71	51 Sb 121.75	52 Te 127.60	53 I 126.905	54 Xe 131.29
6	55 Cs 132.905	56 Ba 137.33	★	72 Hf 178.49	73 Ta 180.948	74 W 183.85	75 Re 186.207	76 Os 190.2	77 Ir 192.22	78 Pt 195.08	79 Au 196.967	80 Hg 200.59	81 Tl 204.383	82 Pb 207.2	83 Bi 208.980	84 Po (209)	85 At (210)	86 Rn (222)
7	87 Fr (223)	88 Ra 226.025	▲	104 Und (261)	105 Unp (262)	106 Unh (263)	107 Uns (264)	108 Uno (265)	109 Une (266)	110 Uun (267)	111 Uuu (272)	112 Uub	113 Uut	114 Uuq	115 Uup	116 Uuh	117 Uus	118 Uuo

TRANSITION ELEMENTS

6 ★ Lanthanide Series (RARE EARTH)

7 ▲ Actinide Series (RARE EARTH)

57 La 138.906	58 Ce 140.12	59 Pr 140.908	60 Nd 144.24	61 Pm (145)	62 Sm 150.36	63 Eu 151.96	64 Gd 157.25	65 Tb 158.925	66 Dy 162.50	67 Ho 164.930	68 Er 167.26	69 Tm 168.934	70 Yb 173.04	71 Lu 174.967
89 Ac 227.028	90 Th 232.038	91 Pa 231.036	92 U 238.029	93 Np 237.048	94 Pu (244)	95 Am (243)	96 Cm (247)	97 Bk (247)	98 Cf (251)	99 Es (252)	100 Fm (257)	101 Md (258)	102 No (259)	103 Lr (260)

which contains 2 electrons. For most of the fifteen lanthanides, the fifth or "O" shell contains 8 electrons (with three exceptions). It is the fourth or "N" shell (third from the outside) that becomes the defining shell in which electrons are added as each element increases in **atomic number**. The appendix "Electron Configuration of the Atoms" depicts how the $4f$ suborbital of the N shell, and not the outer shell, gains electrons.

The elements in the lanthanide series are soft, silver-like metals that form **trivalent** compounds. As metals, they are reactive, **tarnish** when exposed to the atmosphere, and react with most nonmetals when heated. Europium (Er^{68}) is the most reactive. The size and hardness of the elements increase from left to right in the series.

The ores from which rare-Earth elements are extracted are monazite, bastnasite, and oxides of yttrium, and related fluorocarbonate minerals. These ores are found in South Africa, Australia, South America, India, and in the United States in California, Florida, and the Carolinas. Several of the rare-Earth elements are also produced as **fission** by-products during the decay of the radioactive elements uranium and plutonium. The elements of the lanthanide series that have an *even* atomic number are much more abundant than are those of the series that have an *odd* atomic number.

A commercial mix of several of the rare-Earth elements is called didymium (Di). It is not an element, but it is used to name mixtures of **oxides** and salts of most of the rare-Earth elements that are extracted from the ore monazite. Another unique substance, called **misch metal**, is an alloy of iron and several rare-Earth elements (La, Ce, and Pr). It is **pyrophoric**, which means it sparks when scratched. This is why it is used for cigarette lighter flints.

Several rare-Earth elements are used as **control rods** in **nuclear reactors**.

LANTHANUM (Rare-Earth Elements)
SYMBOL: La   **ATOMIC NUMBER:** 57   **PERIOD:** 6

COMMON VALENCE: 3 **ATOMIC WEIGHT:** 138.91 **NATURAL STATE:** Solid **COMMON ISOTOPES:** There are 2 stable isotopes of lanthanum: lanthanum-138 and lanthanum-139. **PROPERTIES:** Lanthanum is a whitish/iron-colored metallic element that **oxidizes** in air at room temperature. Density: 6.18; melting point: 920°C; boiling point: 3454°C.

CHARACTERISTICS
Lanthanum metal is **malleable** and **ductile**; i.e., it can be worked and formed into shapes.

It rapidly corrodes in air and decomposes in water, producing H_2 and La_2O_3.

Lanthanum is considered the most basic (alkaline) of the rare-Earth elements.

ABUNDANCE AND SOURCE

Lanthanum is the twenty-eighth most abundant element on Earth. This so-called "rare-Earth" is not really rare. Like the other elements in the lanthanide series, lanthanum is difficult to separate and identify from its ores because of the similarity with other rare-Earths.

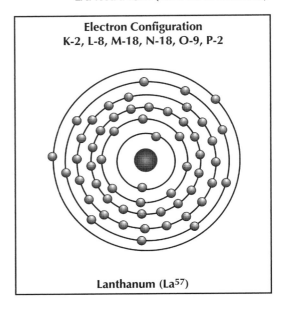

Electron Configuration
K-2, L-8, M-18, N-18, O-9, P-2

Lanthanum (La⁵⁷)

Lanthanum is the second most abundant element of the rare-Earth group. It is usually found with other rare-Earths. Lanthanum is dissolved out of the ores monazite, apatite, and bastnasite by using sulfuric acid. It is also found is some calcite and fluorite minerals. The ores are found in South Africa, Australia, Brazil, and India, and in California, Florida, and the Carolinas in the United States.

Lanthanum is one of the radioactive end products of **fission** decay (giving off particles) of uranium, thorium, and plutonium.

HISTORY

Johan Gadolin (1760–1852) studied some minerals that were different from others known at that time. Since they were different from the common "Earth elements" but were all very similar to each other, he named them "rare-Earth elements." He was not able to separate or identify them.

Carl Gustaf Mosander (1797–1858), while studying the element cerium, found a new element which he named *lanthanum* for the Greek word *lanthano*, which means "hidden" or "concealed." He is given credit for the discovery of lanthanum.

COMMON USES

Lanthanum is used to make glass with a high refractive index which is used for quality lenses in scientific instruments, cameras, and so forth.

Lanthanum, when mixed with iron and several other rare-Earths, becomes a **misch metal** alloy, which is used to make flints for cigarette lighters. Misch metal has the ability to spark when scratched with a sharp object, such as the grooved wheel of a lighter.

Lanthanum is used for electronic instruments, as a rocket fuel, and as a **reducing agent**. It is also used in **catalytic converters**.

EXAMPLES OF COMMON COMPOUNDS

Lanthanum arsenide (LaAs)—Used as a binary **semiconductor**. It is very toxic.

Lanthanum fluoride (LaF_3)—A white powder used to coat the inside of phosphorous lamps and lasers.

Lanthanum oxide (La_2O_3)—Used to make optical glass, ceramics, carbon-arc electrodes, and **fluorescent** lamps.

HAZARDS

In powder form, lanthanum will ignite spontaneously. If ingested, it can cause liver damage and prevent blood from clotting.

CERIUM (Rare-Earth Elements)
SYMBOL: Ce   **ATOMIC NUMBER:** 58   **PERIOD:** 6

COMMON VALENCE: 3 and 4 **ATOMIC WEIGHT:** 140.12 **NATURAL STATE:** Solid   **COMMON ISOTOPES:** There are 4 stable isotopes, plus several radioactive isotopes. Cerium-141 with a half-life of 21.5 days is derived from cerium-140 by the capture of a neutron. Other isotopes are: cerium-136, cerium-138, and cerium-142.   **PROPERTIES:** Cerium is a grayish/iron-colored, very reactive metallic element that is attacked by both acids and alkalies. It is soft and **malleable**. Density: 6.78; melting point: 795°C; boiling point: 3257°C.

CHARACTERISTICS

Cerium will rapidly **oxidize** in moist air at room temperature.
Cerium is the second most reactive of the lanthanide series of rare-Earths.
It decomposes in water to form hydrogen and hydroxides.

ABUNDANCE AND SOURCE

Cerium is the twenty fifth-most abundant element on the Earth. It is also the most abundant of the rare-Earth metals in the lanthanide series. It is found in ores with other rare-Earths.

The ore bastnasite is found in California and New Mexico. The ore monazite is found in South Africa, India, Brazil, and in the beach sands of Florida. The ore cerite is found in Sweden.

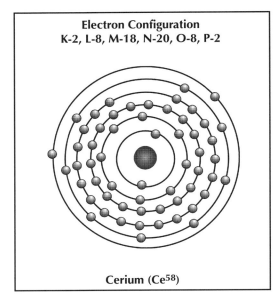

Electron Configuration
K-2, L-8, M-18, N-20, O-8, P-2

Cerium (Ce⁵⁸)

HISTORY

Martin Heinrich Klaproth (1743–1817), Jöns Jakob Berzelius (1779–1848) and Wilhelm Hisinger (1766–1852) were all given credit for the discovery of cerium in 1803. They named it *cerium* after the newly discovered asteroid Ceres.

It was not until 1875 that W. F. Hillebrand and T. H. Norton purified the metal cerium.

COMMON USES

Cerium is used to make special alloys for jet engines, solid-state instruments, and rocket propellant.

Cerium is **pyrophoric**, which means it can produce sparks by friction. This quality makes it ideal as a **misch metal** alloy used to make flints for cigarette lighters and similar ignition devices.

It is also used in **catalytic converters** and as a **catalyst** in the petroleum industry.

EXAMPLES OF COMMON COMPOUNDS

Cerium-141—Used as a radioactive source in medicine.

Cerous fluoride (CeF_3)—Used to increase brightness in carbon arc lamps.

Cerous nitrate [$Ce(NO_3)_3$]—Used to separate cerium from the other rare-Earths.

HAZARDS

Cerium will ignite when heated. Cerium compounds are toxic.

PRASEODYMIUM (Rare-Earth Elements)
SYMBOL: Pr   ATOMIC NUMBER: 59   PERIOD: 6

COMMON VALENCE: 3 and 4 **ATOMIC WEIGHT:** 140.91 **NATURAL STATE:** Solid **COMMON ISOTOPES:** The stable form, praseodymium-140.9077, makes up 100% of the naturally occurring element. **PROPERTIES:** Depending on how it is separated from its ores, praseodymium can be a silver-yellowish metal or a black powder. It is a soft, **ductile**, **malleable** metallic element. Density: 6.8; melting point: 930°C; boiling point: 3200°C.

CHARACTERISTICS

As a metal, Pr is **hygroscopic** (absorbs water) and **tarnishes** in the atmosphere.

It will react with water to liberate hydrogen.

It is soluble in acids and forms greenish-colored salts.

ABUNDANCE AND SOURCE

Praseodymium is the thirty-ninth most abundant element on Earth. As one of the rare-Earths, praseodymium is not very abundant. It is found mixed with several of the other rare-Earths in the ores of monazite, cerite, bastnasite, and allanite.

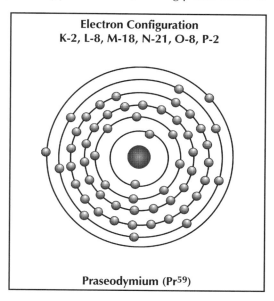

Electron Configuration
K-2, L-8, M-18, N-21, O-8, P-2

Praseodymium (Pr59)

Praseodymium is also a stable by-product of the **fission** of some other elements, such as uranium.

HISTORY

At first praseodymium was called *didymium*, which is Greek for "twin," because it was always found with another rare-Earth element.

In 1885 Carl Auer Baron von Welsbach (1858–1929) separated **oxides** of the two similar elements. He named one *praseodymium* from the Greek word *prasios*, which means "leek green," and which he referred to as the "green twin." He named the other similar element *neodymium*, which is derived from "new + dymium," and means "new twin."

COMMON USES

The main use of praseodymium is as a pigment to color glass and ceramics. It is also used for carbon **electrodes** for arc lamps.

It is used to make lenses and goggles for situations in which strong yellow light must be filtered, such as when welding.

EXAMPLES OF COMMON COMPOUNDS

There are not many compounds.

Praseodymium oxalate [$Pr_2(C_2O_4)_3$]—A green powder used to color ceramics.

Praseodymium oxide (Pr_2O_3)—A yellowish powder used as a glass and ceramic pigment.

HAZARDS

Praseodymium may explode as it liberates hydrogen from water.

❖   ❖   ❖

NEODYMIUM (Rare-Earth Elements)
SYMBOL: Nd   **ATOMIC NUMBER:** 60   **PERIOD:** 6

COMMON VALENCE: 3 **ATOMIC WEIGHT:** 144.24 **NATURAL STATE:** Solid **COMMON ISOTOPES:** Neodymium has 7 isotopes. **PROPERTIES:** Neodymium is a silver-yellow metal in its elemental form. It can form several **trivalent** salts which exhibit reddish or violet-like colors. Density: 7.0; melting point: 1024°C, boiling point: 3000°C.

CHARACTERISTICS

Neodymium is a soft metal of the lanthanide, rare-Earth series.

It is **malleable**, i.e., can be worked, cut, and machined.

It oxidizes with moist air and must be stored under oil.

Neodymium reacts with water to release hydrogen.

ABUNDANCE AND SOURCE

Neodymium is the twenty-seventh most abundant element on Earth, but it is not as abundant as some of the other rare-Earths.

Neodymium is separated from other rare-Earths

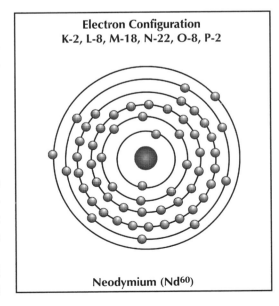

Electron Configuration
K-2, L-8, M-18, N-22, O-8, P-2

Neodymium (Nd[60])

found in the ores monazite, bastnasite, and allanite by heating with sulfuric acid. It can be formed into rods, sheets, powder, or ingots.

HISTORY

In 1885 Carl Auer Baron von Welsbach (1858–1929) discovered neodymium, which formed blue salts when he separated it from the very similar rare-Earth, praseodymium. Since neodymium was so similar to praseodymium, he gave it the name *neodymium*, which means "new twin."

COMMON USES

Neodymium is used in the electronics industry, and to make blue-colored glass for astronomical and laboratory observation instruments since it absorbs the yellow "sodium" line in the visible light **spectrum**.

It is used to make alloys with other metals, and to make the purple glass for lasers and for coating color TV tubes. It is also used to color eyeglasses.

Neodymium, along with several other rare-Earths, is also used to make **misch metal** alloys with iron, which are used to make flints for cigarette lighters.

EXAMPLES OF COMMON COMPOUNDS

Neodymium oxide (Nd_2O_3)—A light-blue powder used to color glass and ceramics and to make color TV tubes.

Neodymium fluoride (NdF_3)—A pink powder used as a pigment.

HAZARDS

Many of the compounds (salts) of neodymium are skin irritants and toxic if inhaled or ingested. Some are explosive, e.g., neodymium nitrate [$Nd(NO_3)_3$], which is used as a pink pigment.

PROMETHIUM (Rare-Earth Elements)
SYMBOL: Pm   **ATOMIC NUMBER:** 61   **PERIOD:** 6

COMMON VALENCE: 2   **ATOMIC WEIGHT:** 147   **NATURAL STATE:** Solid
COMMON ISOTOPES: All forms of promethium are artificially produced and radioactive. Promethium-145 has a half-life of 18 years. Promethium-147 has a half-life of 2.6 years. **PROPERTIES:** Promethium is a silvery-white, very radioactive metal which is recovered as a by-product of uranium fission. Promethium-147 is the only isotope available. Density: 7.2; melting point: 1160°C; boiling point: 3074°C.

CHARACTERISTICS

Promethium was the missing element of the lanthanide rare-earth series.

It was predicted by the Periodic Table, but elusive to identify because of its similarity to other rare-Earths and its radioactivity.

ABUNDANCE AND SOURCE

Promethium is not found in nature. All of it is man-made, and it is very rare. It is found only as a "**transmuted**" decay by-product element from the **fission** of radioactive uranium.

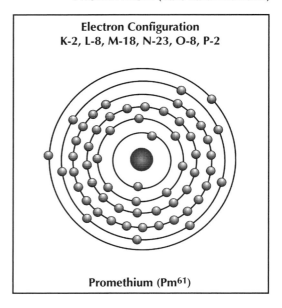

Electron Configuration
K-2, L-8, M-18, N-23, O-8, P-2

Promethium (Pm⁶¹)

Promethium-147 is found in **nuclear** reactors as a **fission** decay product.

HISTORY

The identification of promethium-147 was verified by using the mass **spectrometer**, which analyzes light rays. Several radioisotopes of promethium were identified by a **spectrograph**, which is something like a photograph depicting the lines represented by different wavelengths of light rays.

It was named *promethium* after the Titan Prometheus, who in Greek mythology stole fire from the gods to give it to mankind. The name was suggested by Grace Mary Coryell, wife of Charles DuBois Coryell (1912–1995), the scientist who led a team working with the by-products of the decay of uranium. The credit for the separation and identification of promethium was given to J. A. Marinsky and L. E. Glendenin in 1945.

COMMON USES

A recent use is for nuclear-powered batteries to provide electricity for spacecraft.

Promethium-147 is used to make **luminescent** paint for watch dials, and as a source of **beta rays** for thickness gauges.

EXAMPLES OF COMMON COMPOUNDS

Over 30 compounds are known, but few are available.

HAZARDS

Promethium is an extremely strong radiation hazard.

Safety precautions must be used when working with promethium and its radioactive isotopes.

SAMARIUM (Rare-Earth Elements)

SYMBOL: Sm   **ATOMIC NUMBER:** 62   **PERIOD:** 6

COMMON VALENCE: 2 and 3 **ATOMIC WEIGHT:** 150.35 **NATURAL STATE:** Solid **COMMON ISOTOPES:** Samarium has 7 stable natural isotopes. Three radioactive isotopes emit **alpha particles**. They are samarium-147, samarium-148, and samarium-149. **PROPERTIES:** Samarium is a hard, brittle, silvery-white metal that forms a yellow **oxide** coating when exposed to the atmosphere. Density: 7.53; melting point: 1072°C; boiling point: 1900°C.

CHARACTERISTICS

Samarium will ignite at the rather low temperature of 150°C.

It is an active **reducing agent** and will liberate hydrogen from water.

Samarium has the capacity to absorb neutrons in **nuclear reactors**.

ABUNDANCE AND SOURCE

Samarium is the fortieth most abundant element on Earth. It is found in minerals such as gadolinite, cerite, monazite, and samarskite in South Africa, South America, Australia, and in the southeastern states of the United States. It is also found in bastnasite ores in California. Samarium is also a by-product of **fission** in nuclear reactors.

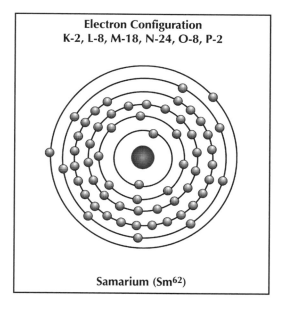

Electron Configuration
K-2, L-8, M-18, N-24, O-8, P-2

Samarium (Sm^{62})

HISTORY

In 1879 Paul Emile Lecoq de Boisbaudran (1838–1912) identified a black **oxide** that was contaminated by two other rare-Earth minerals, gadolinia and europia. This oxide, once separated, turned out to be a new rare-Earth. He named it *samarium* for the mineral samarskite, which was named for Col. von Samarski, a Russian mine official. Samarskite ore is where didymia is found. Didymia (twins) is the original name given to a combination of the two rare-Earths praseodymium and neodymium, before they were separated. Samarium was first identified by **spectroscopy**.

COMMON USES

Samarium is used as a neutron absorber in nuclear reactors. It is also used to make **lasers**, magnets, and for **metallurgical** research.

Samarium is used to make special magnets for computer hard disks and related devices, and to make **misch metal** alloys for flints used in cigarette lighters.

EXAMPLES OF COMMON COMPOUNDS

Samarium oxide (Sm_2O_3)—Used as a catalyst to speed up **dehydrogenation** reactions, and as a **neutron absorber**. It will also absorb carbon dioxide from the air.

HAZARDS

The salts of samarium are toxic if ingested. When placed in water, samarium liberates hydrogen, which may explode.

EUROPIUM (Rare-Earth Elements)
SYMBOL: Eu   **ATOMIC NUMBER:** 63   **PERIOD:** 6

COMMON VALENCE: 2 and 3 **ATOMIC WEIGHT:** 151.96 **NATURAL STATE:** Solid **COMMON ISOTOPES:** Two stable, naturally occurring isotopes of europium are europium-151 and europium-153. **PROPERTIES:** Europium is a steel-gray, very soft metal that is difficult to prepare, but quite **ductile** and **malleable**, i.e., can be worked and formed into many shapes. Europium is the last rare-Earth of what is sometimes referred to as the cerium subgroup of the lanthanide series. Density: 5.24; melting point: 826°C; boiling point: 1490°C.

CHARACTERISTICS

The chemistry of europium is more closely related to that of the elements calcium, strontium, and barium of Group 2 than to the rare-Earths.

Europium **oxidizes** rapidly in air and will react with water to release hydrogen.

ABUNDANCE AND SOURCE

Europium is the fiftieth most abundant element on Earth. It is found in most of the rare-Earth ores, such as monazite, bastnasite, cerite, and allanite.

HISTORY

In 1901 Eugene-Anatole Demarcay (1852–1904) discovered europium **spectroscopically** and was the first to separate it from the rare-Earth samarium. He named it after the continent of Europe.

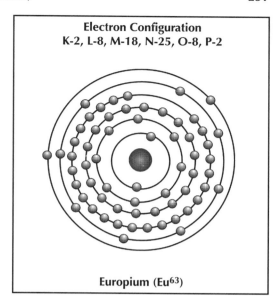

Electron Configuration
K-2, L-8, M-18, N-25, O-8, P-2

Europium (Eu^{63})

COMMON USES

Europium is used as a **neutron absorber** in **nuclear reactors**, to add red color to TV tubes, and as an addition to the glue on postage stamps so they can be read by electronic sorting machines in U.S. post offices.

EXAMPLES OF COMMON COMPOUNDS

Europium oxide (Eu_2O_3)—Used to make infrared-sensitive **phosphors**, and as nuclear reactor **control rods**.

HAZARDS

Europium is very reactive and, in powder form, may burst into flames spontaneously at room temperature. Most of the salts of europium are toxic when inhaled and ingested.

GADOLINIUM (Rare-Earth Elements)
SYMBOL: Gd   **ATOMIC NUMBER:** 64   **PERIOD:** 6

COMMON VALENCE: 3 **ATOMIC WEIGHT:** 157.25 **NATURAL STATE:** Solid **COMMON ISOTOPES:** Gadolinium has 7 natural isotopes. **PROPERTIES:** Gadolinium is silvery with a metallic luster. It is the first of what is sometimes referred to as the yttrium subgroup (lanthanide series) of rare-Earths. Gadolinium is combustible and burns in air to form gadolinium oxide (Gd_2O_3), which is white, while its salts are colorless. Density: 7.87; melting point: 1312°C; boiling point: 3000°C.

CHARACTERISTICS

Gadolinium, unlike most of the rare-Earths in the yttrium subgroup, reacts slowly with water.

It exhibits strong magnetism at low temperatures.

Gadolinium has the highest **neutron absorption** capability (neutron-stopping power) of any known element.

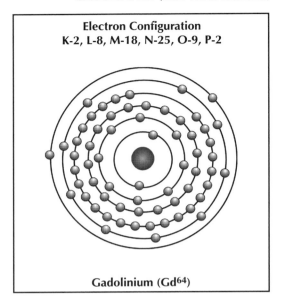

Electron Configuration
K-2, L-8, M-18, N-25, O-9, P-2

Gadolinium (Gd⁶⁴)

ABUNDANCE AND SOURCE

Gadolinium is the forty-first most abundant element on Earth. It is found along with other rare-Earths in most of the ores that produce the rare-Earths, e.g., monazite, bastnasite, cerite, and allanite. It is also produced by nuclear **fission**.

HISTORY

In 1878 Jean-Charles Gallissard de Marignac (1817–1894) discovered one of the "missing" rare-Earths, which he named *ytterbium*. Two years later Marignac discovered another, which was named *gadolinium* for Johan Gadolin (1760–1852), the "father of the rare-Earths." Gadolin was given this honor because he discovered the first rare-Earth and, in general, described the similar chemical properties of all the rare-Earths.

COMMON USES

Gadolinium's main use is based on its ability to absorb neutrons. This makes it ideal as a neutron-shielding metal.

It is also used as a **catalyst** to speed up **chemical reactions**, as a component of microwave filters, and to activate **phosphor** compounds in TV screens. It is also used in high-temperature furnaces.

Gadolinium is **paramagnetic** at normal temperatures (weaker than **ferromagnetic**) and becomes strongly ferromagnetic at very cold temperatures.

EXAMPLES OF COMMON COMPOUNDS

Gadolinium oxide (Gd_2O_3)—Used as neutron shields, to coat **filaments** of electrical devices, to make special glass, and as a TV screen phosphor activator.

Gadolinium sulfate [$Gd_2(SO_4)_3$]—Used in **thermoelectric** devices that convert heat to electricity, and in **cryogenic** (very low temperature) research.

HAZARDS

The halogens of gadolinium are very toxic. Gadolinium nitrate is explosive. As with most of the rare-Earths, care should be taken not to inhale fumes or ingest particles of gadolinium.

❖ ❖ ❖

TERBIUM (Rare-Earth Elements)
SYMBOL: Tb   **ATOMIC NUMBER:** 65   **PERIOD:** 6

COMMON VALENCE: 3 and 4 **ATOMIC WEIGHT:** 158.925 **NATURAL STATE:** Solid **COMMON ISOTOPES:** No stable isotopes. Terbium-159 makes up 100% of the natural element. **PROPERTIES:** Terbium is a very reactive, silver-gray metallic element found in the yttrium subgroup (lanthanide series) of the rare-Earths. Density: 8.332; melting point: 1356°C; boiling point: 2800°C.

CHARACTERISTICS

Terbium reacts slowly in air at room temperature but will burn in air at higher temperatures.

It is **ductile** and **malleable**.

Tb must be handled in an inert atmosphere.

ABUNDANCE AND SOURCE

Terbium is the fifty-seventh most abundant element on Earth. It is not a very abundant free metal and is usually found in compound form. It is found in the ores of other rare-Earths, such as monazite, bastnasite, and allanite.

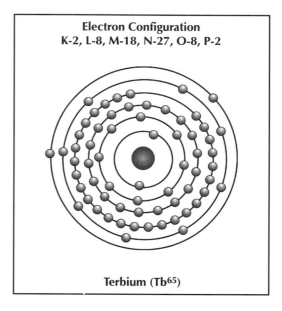

Electron Configuration
K-2, L-8, M-18, N-27, O-8, P-2

Terbium (Tb⁶⁵)

HISTORY

Carl Gustaf Mosander (1797–1858) discovered several rare-Earths. Among those he identified was terbium, which was named for one of the syllables of Ytterby, a stone quarry in Sweden where many of the rare-Earths were first found.

COMMON USES

Terbium has limited uses. One is to coat solid-state devices; another is to activate the **phosphors** for TV picture tubes. It is also used in **control rods** of **nuclear reactors**.

EXAMPLES OF COMMON COMPOUNDS

Terbium nitrate [Tb(NO₃)₃]—A fire risk when in contact with organic substances. Used as an activator for solid-state devices.

Terbium oxide (Tb₂O₃)—A brown powder that absorbs carbon dioxide.

HAZARDS

The halogens of terbium are strong irritants. Most of the compounds are toxic and some are explosive. A vacuum or inert gas environment must be maintained when working with the metal because of its strong **oxidation** properties.

❖   ❖   ❖

DYSPROSIUM (Rare-Earth Elements)
SYMBOL: Dy   **ATOMIC NUMBER:** 66   **PERIOD:** 6

COMMON VALENCE: 3   **ATOMIC WEIGHT:** 162.50   **NATURAL STATE:** Solid

COMMON ISOTOPES: There are 6 stable isotopes of dysprosium in addition to the main, naturally occurring element. **PROPERTIES:** Dysprosium is a shiny, soft metal that does not react with air at room temperature. Density: 8.54; melting point: 1407°C; boiling point: 2330°C.

CHARACTERISTICS

Although dysprosium does not react with moist air at low temperatures, it does react with water and the halogens at high temperatures, and it is soluble in weak acids.

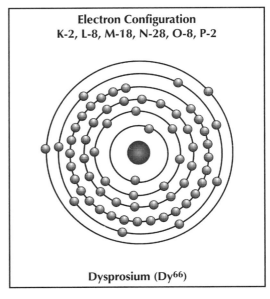

Electron Configuration
K-2, L-8, M-18, N-28, O-8, P-2

Dysprosium (Dy⁶⁶)

Dysprosium is strongly magnetic at very low temperatures.

ABUNDANCE AND SOURCE

Dysprosium is the fiftieth most abundant element. It is a metallic element usually found as an **oxide**. It is found, as are most other rare-Earths, in ores of monazite and allanite, in Africa, South America, Australia, and southeastern United States. It is also found in the ore bastnasite, in California.

HISTORY

In 1886 Paul-Emile Lecoq de Boisbaudran (1838–1912) was working with his newly discovered rare-Earth holmium, when he was able to gather a small amount of another new rare-Earth. He called it *dysprosium* from the Greek word *dusprositos*, which means "difficult to approach" or "hard to get at."

COMMON USES

Dysprosium is used as an alloy to make **control rods** for **nuclear reactors**. It is also used as a **fluorescence** activator for the **phosphors** used to coat TV screens.

Dysprosium makes excellent magnetic alloys when combined with other metals, such as iron and nickel.

EXAMPLES OF COMMON COMPOUNDS

Dysprosium oxide (Dy_2O_3)—A white powder that is more magnetic than iron oxide; when dissolved in acid, it becomes yellow-green.

HAZARDS

Dysprosium nitrate [$Dy(NO_3)_3$] is a strong **oxidizing agent** and will ignite when in contact with organic material. Most of the dysprosium salts are toxic if ingested or inhaled.

HOLMIUM (Rare-Earth Elements)
SYMBOL: Ho   **ATOMIC NUMBER:** 67   **PERIOD:** 6

COMMON VALENCE: 3 **ATOMIC WEIGHT:** 164.903 **NATURAL STATE:** Solid **COMMON ISOTOPES:** There are no stable isotopes. Holmium-165 makes up 100% of the naturally occurring element. **PROPERTIES:** Holmium is a crystal-like, solid rare-Earth with a metallic luster. Holmium is found in the yttrium subgroup (lanthanide series) of the Periodic Table. Density: 8.803; melting point: 1470°C; boiling point: 2720°C.

CHARACTERISTICS

Holmium reacts slowly with water and will dissolve in weak acids.

Holmium has one of the highest magnetic properties of any substance.

It is soft, **ductile**, and **malleable**.

ABUNDANCE AND SOURCE

Holmium is the fifty-fifth most abundant element and it is one of the least abundant lanthanide metals. It is found in gadolinite and monazite ores, in South Africa, Australia, Florida, and the Carolinas in the United States.

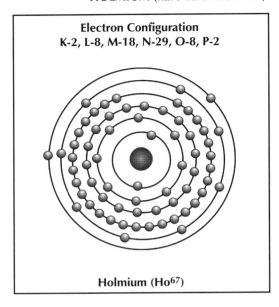

Electron Configuration
K-2, L-8, M-18, N-29, O-8, P-2

Holmium (Ho^{67})

HISTORY

In 1878 J. Louis Soret (1827–1890) and Marc Delafontaine had **spectroscopic** evidence of the element holmium, but it was contaminated by the rare-Earth mineral dysprosia. They did not get credit for the discovery because they could not isolate holmium.

In 1879 holmium was discovered, independently, by Per Theodor Cleve (1840–1905), who did separate the elements. Cleve, who received credit for the discovery, named *holmium* for the Latin word *holmia*, which means "Stockholm," a city in his native country, Sweden.

COMMON USES

Holmium's oxides have little commercial value.

It is used as **filaments** in vacuum tubes and in **electro-chemistry.** It is also used to help identify the atomic weights of elements by **spectroscopy**, which identifies the unique lines produced by each element when viewed through a spectroscope.

It is used as a **neutron absorber** in reactors.

EXAMPLES OF COMMON COMPOUNDS

Holmium oxide (Ho$_2$O$_3$)—A pale yellow solid used as a special catalyst to speed up chemical reactions and as a **refractory material** to line laboratory and industrial ovens.

HAZARDS
The hazards of holmium are the same as for other rare-Earths. The fumes and dust of both the elemental form and its compounds should not be inhaled or ingested.

❖ ❖ ❖

ERBIUM (Rare-Earth Elements)
SYMBOL: Er   ATOMIC NUMBER: 68   PERIOD: 6

COMMON VALENCE: 3 **ATOMIC WEIGHT:** 167.26 **NATURAL STATE:** Solid **COMMON ISOTOPES:** There are 6 stable isotopes of erbium that make up the natural element's average atomic weight of 167.26. **PROPERTIES:** Erbium is a soft, **malleable** metal with a silvery metallic luster. It is one of the rare-Earths of the yttrium subgroup of the lanthanide series. Density: 9.16; melting point: 1522°C; boiling point: 2550°C.

CHARACTERISTICS
Erbium is soft and can easily be worked into different shapes.

It is insoluble in water but is soluble in acids. Its salts range from pink to red.

Erbium exhibits high resistance to electricity.

ABUNDANCE AND SOURCE
Erbium is the forty-fourth most abundant element on Earth. It is found in rare-Earth ores such as monazite, gadolinite, and bastnasite. It is also a by-product of nuclear fission of uranium.

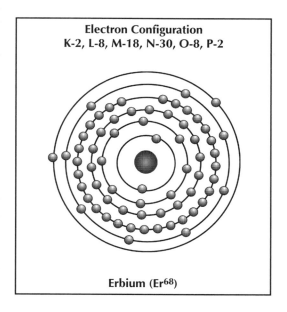

**Electron Configuration
K-2, L-8, M-18, N-30, O-8, P-2**

Erbium (Er⁶⁸)

HISTORY
In 1843 Carl Gustaf Mosander (1797–1858) isolated several new elements from the mineral gadolinite. One he named *erbium* for one of the syllables of the name Ytterby, which is the name of the quarry in Sweden from which these minerals were first found.

COMMON USES

Erbium is used to make nuclear **control rods** for fission reactors and to make special alloys with other metals, especially with vanadium to make spring steel. It is also used to make **lasers** that will operate at normal room temperature.

EXAMPLES OF COMMON COMPOUNDS

Erbium oxide (Er_2O_3)—A pink powder that absorbs carbon dioxide. Used to make **infrared** absorbing glass.

HAZARDS

Erbium nitrate [$Er(NO_3)_3$] may explode upon being shocked or at high temperatures. As with other rare-Earths, erbium and its compounds should be handled with care because they can be toxic.

❖ ❖ ❖

THULIUM (Rare-Earth Elements)
SYMBOL: Tm   **ATOMIC NUMBER:** 69   **PERIOD:** 6

COMMON VALENCE: 3   **ATOMIC WEIGHT:** 168.9342   **NATURAL STATE:** Solid   **COMMON ISOTOPES:** Thulium-169 makes up 100% of the natural element. There are no other stable isotopes. **PROPERTIES:** Thulium is a rare-Earth element with a metallic luster. It is a soft, **malleable**, **ductile** silver/gray metallic element. Thulium is located in the yttrium subgroup of the lanthanide series of the Periodic Table. Density: 9.318; melting point: 1550°C; boiling point: 1730°C.

CHARACTERISTICS

Thulium reacts slowly with water and is soluble in weak acids.

When combining with oxygen and the halogens, thulium produces salts that are a pale-green color.

ABUNDANCE AND SOURCE

Thulium is the sixty-first most abundant element on Earth and is usually found

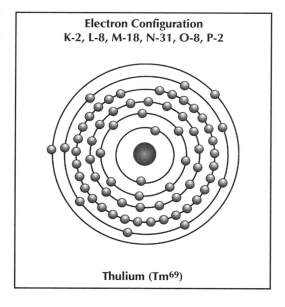

Electron Configuration
K-2, L-8, M-18, N-31, O-8, P-2

Thulium (Tm^{69})

as an oxide along with the other rare-Earths in the ores of the yttrium subgroup, e.g., monazite and bastnasite. It also occurs as a by-product of nuclear **fission**.

HISTORY
In 1879 Per Theodor Cleve (1840–1905) discovered two new rare-Earths, holmium and thulium. He named *thulium* after the Greek word *Thule*, which in legends represented the most northern region inhabited by humans, now known as Scandinavia.

COMMON USES
Thulium-170, which is radioactive thulium produced in **nuclear reactors**, is used as a small, portable X-ray source.

It is also used to make magnetic alloys.

EXAMPLES OF COMMON COMPOUNDS
Thulium oxide (Tm_2O_3)—A heavy powder, slightly whitish-green but turns slightly reddish, then yellowish on heating. It is the main source of thulium metal.

HAZARDS
The dust and powder of thulium are explosive and toxic if inhaled or ingested. As with all radioactive elements, thulium can cause radiation poisoning.

YTTERBIUM (Rare-Earth Elements)
SYMBOL: Yb   **ATOMIC NUMBER:** 70   **PERIOD:** 6

COMMON VALENCE: 2 and 3 **ATOMIC WEIGHT:** 173.04 **NATURAL STATE:** Solid   **COMMON ISOTOPES:** There are 7 stable, naturally occurring isotopes of ytterbium. **PROPERTIES:** Ytterbium is a soft, **ductile**, **malleable** rare-Earth element with a metallic luster, which is not very reactive. It is located in the yttrium subgroup of the lanthanide series. Density: 7.0; melting point: 824°C; boiling point: 1400°C.

CHARACTERISTICS
Ytterbium reacts slowly with water and is soluble in weak acids.

The salts of ytterbium are **paramagnetic**, which exhibit weaker magnetic fields than do iron magnets.

ABUNDANCE AND SOURCE
Ytterbium is the forty-third most abundant element and is found in ores along with other rare-Earths, first found in the Ytterby quarry of Sweden. These ores are xenotime, euxenite, gadolinite, and monazite.

Ytterbium is also found as **fission** decay products of fission **nuclear reactions**.

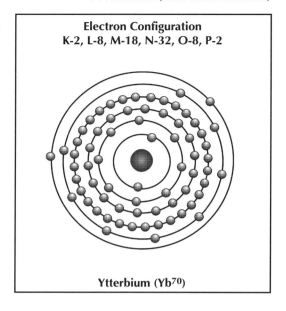

Electron Configuration
K-2, L-8, M-18, N-32, O-8, P-2

Ytterbium (Yb^{70})

HISTORY

So many scientists were looking for and identifying the rare-Earths during the mid- to late 1800s that it is not always easy to determine who was the actual discoverer of ytterbium.

Lars Fredrik Nilson (1840–1899) is given credit for discovering ytterbium, but other sources claim Carl Gustaf Mosander (1797–1858) as the discoverer because he separated several rare-Earths, including ytterbium, from ores. The element was named *ytterbium* after Ytterby, which is the name of the town in Sweden where the feldspar quarry is located, and where ytterbium was found.

Other sources claim that Georges Urbain (1872–1938) and Carl Auer Baron von Welsbach (1858–1929) independently discovered ytterbium. And J.-C. G. de Marignac (1817–1894) is also credited with the discovery of ytterbium in 1878 by some authorities.

COMMON USES

There are not many commercial uses for ytterbium.

Radioactive ytterbium is used as a small portable X-ray source. Ytterbium metal and its oxide are used to make **lasers**, and in chemical research. Ytterbium is also used to make synthetic gemstones.

EXAMPLES OF COMMON COMPOUNDS

Ytterbium oxide (Yb_2O_3)—Used to make special alloys, ceramics, and glass. Also used for carbon arc lamp rods to produce very bright light sources.

HAZARDS

Ytterbium dust and powder can explode and may be toxic if inhaled. Ytterbium arsenate is a poison.

❖ ❖ ❖

LUTETIUM (Rare-Earth Elements)
SYMBOL: Lu   **ATOMIC NUMBER:** 71   **PERIOD:** 6

COMMON VALENCE: 3 **ATOMIC WEIGHT:** 174.97 **NATURAL STATE:** Solid **COMMON ISOTOPES:** Lutetium-175 is the only stable natural form of the element. Lutetium-176 is a **beta-emitting** radioisotope with a half-life of 2.2×10^{10} years (22 billion years). **PROPERTIES:** Lutetium is a soft, workable, rare-Earth element with a metallic luster. It is the heaviest and last rare-earth in the lanthanide series in the Periodic Table. Density: 10.849; melting point: 1652°C; boiling point: 3327°C.

CHARACTERISTICS

Lutetium reacts slowly with water and is soluble in weak acids.

Lutetium's crystals exhibit strong magnetic properties which are important to the study of magnetism.

ABUNDANCE AND SOURCE

Lutetium is the sixtieth most abundant element on Earth. It is one of the rarest of the lanthanide elements. Lutetium is found in the ore monazite, in India, Australia, Africa, South America, and the United States. It also is found as a **fission** product in **nuclear reactors**.

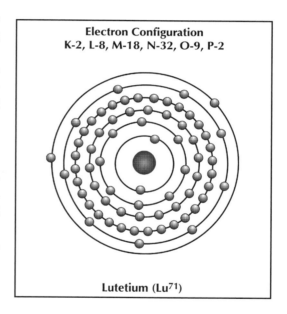

Electron Configuration
K-2, L-8, M-18, N-32, O-9, P-2

Lutetium (Lu⁷¹)

HISTORY

Lutetium was discovered some years after thirteen of the other rare-Earths had already been separated and analyzed. In 1907 Georges Urbain (1872–1938) discovered lutetium, which he named after the Latin word *Lutetia,* the name of the old Roman town that later became Paris, France.

This is another instance where another scientist independently discovered the same element. Carl Auer Baron von Welsbach (1858–1929) announced his discovery at about the same time, but did not get credit.

COMMON USES

The main use of lutetium is in **nuclear reactors**. It is also used to determine the age of meteorites.

Lutetium is used for **cracking** (refining) petroleum.

EXAMPLES OF COMMON COMPOUNDS

Lutetium oxide (Lu_2O_3)—A white solid that is the oxide of lutetium found in monazite ore. It is **hygroscopic** and absorbs carbon dioxide.

Lutetium nitrate [$Lu(NO_3)_3$]—A fire and explosion hazard when heated.

HAZARDS

Lutetium fluoride is a skin irritant and is toxic if inhaled. The dust and powder forms of lutetium and its salts are toxic if inhaled or ingested.

10

ACTINIDE, TRANSURANIC, AND TRANSACTINIDE SERIES

There are three blocks to this series of element. They are all somewhat similar to **homologue** elements in (vertical) Groups of elements just above them. This means they have very similar physical properties, chemical characteristics, and unique structures, and most, but not all, are man-made. They are found in Period 7, starting with Group 3 of the Periodic Table. Both the sizes and radioactivity of the elements in the total **actinide series** increase from left to right, which means the ones on the right side of the series are more prone to spontaneous **fission** (SF) and generally have shorter **half-lives**. Elements in the total actinide series are not the ones with which you will become most familiar. They are not that common or well known, and most of the ones with higher atomic weights (and numbers) exist for only a few minutes or milliseconds. In some references the actinide series is called the second rare-Earth series. All the elements is this series have physical properties and chemical characteristics similar to the element actinium (Ac^{89}). They are very unstable **radioactive** metals, which makes it difficult to study them even though they are all somewhat similar to actinium, as well as their homologue element, which is located just above each actinide in Period 6 of the Periodic Table. As with the lanthanides (Period 6), the elements in the actinide series do not utilize their outer shells as the valence shells that gain or lose electrons. All of the elements in this series have 2 electrons in their outer "Q" shell. Most, but not all, have 9 electrons in their "P" shell. It is usually the fifth or "O" shell that is the

PERIODIC TABLE OF THE ELEMENTS

GROUPS PERIODS	1 IA	2 IIA	3 IIIB	4 IVB	5 VB	6 VIB	7 VIIB	8	9 VIII	10	11 IB	12 IIB	13 IIIA	14 IVA	15 VA	16 VIA	17 VIIA	18 VIIIA
1	1 H 1.0079																	2 He 4.00260
2	3 Li 6.941	4 Be 9.01218											5 B 10.81	6 C 12.011	7 N 14.0067	8 O 15.9994	9 F 18.9984	10 Ne 20.179
3	11 Na 22.9898	12 Mg 24.305											13 Al 26.9815	14 Si 28.0855	15 P 30.9738	16 S 32.066(6)	17 Cl 35.453	18 Ar 39.948
4	19 K 39.0983	20 Ca 40.08	21 Sc 44.9559	22 Ti 47.88	23 V 50.9415	24 Cr 51.996	25 Mn 54.9380	26 Fe 55.847	27 Co 58.9332	28 Ni 58.69	29 Cu 63.546	30 Zn 65.39	31 Ga 69.72	32 Ge 72.59	33 As 74.9216	34 Se 78.96	35 Br 79.904	36 Kr 83.80
5	37 Rb 85.4678	38 Sr 87.62	39 Y 88.9059	40 Zr 91.224	41 Nb 92.9064	42 Mo 95.94	43 Tc (98)	44 Ru 101.07	45 Rh 102.906	46 Pd 106.42	47 Ag 107.868	48 Cd 112.41	49 In 114.82	50 Sn 118.71	51 Sb 121.75	52 Te 127.60	53 I 126.905	54 Xe 131.29
6	55 Cs 132.905	56 Ba 137.33	★	72 Hf 178.49	73 Ta 180.948	74 W 183.85	75 Re 186.207	76 Os 190.2	77 Ir 192.22	78 Pt 195.08	79 Au 196.967	80 Hg 200.59	81 Tl 204.383	82 Pb 207.2	83 Bi 208.980	84 Po (209)	85 At (210)	86 Rn (222)
7	87 Fr (223)	88 Ra 226.025	▲	104 Und (261)	105 Unp (262)	106 Unh (263)	107 Uns (264)	108 Uno (265)	109 Une (266)	110 Uun (267)	111 Uuu (272)	112 Uub (277)	113 Uut	114 Uuq	115 Uup	116 Uuh	117 Uus	118 Uuo

TRANSITION ELEMENTS

6 ★ Lanthanide Series (RARE EARTH)

57 La 138.906	58 Ce 140.12	59 Pr 140.908	60 Nd 144.24	61 Pm (145)	62 Sm 150.36	63 Eu 151.96	64 Gd 157.25	65 Tb 158.925	66 Dy 162.50	67 Ho 164.930	68 Er 167.26	69 Tm 168.934	70 Yb 173.04	71 Lu 174.967

7 ▲ Actinide Series (RARE EARTH)

89 Ac 227.028	90 Th 232.038	91 Pa 231.036	92 U 238.029	93 Np 237.048	94 Pu (244)	95 Am (243)	96 Cm (247)	97 Bk (247)	98 Cf (251)	99 Es (252)	100 Fm (257)	101 Md (258)	102 No (259)	103 Lr (260)

defining shell in which electrons are added as each element increases in **atomic number** (number of protons), from left to right in Period 7. (See the appendix "Electron Configuration of the Atoms."

The actinides represent the first block of the total series which starts with the element actinium (Ac^{89}) and proceeds through uranium (U^{92}). The second block of elements is referred to as the **transuranic** (beyond uranium) elements. This block begins at neptunium (Np^{93}) and proceeds through the element lawrencium (Lr^{103}), which is also, at times, given the symbol Lw. The third block is referred to as the **transactinide** (beyond the actinide) elements and starts with unniliquadium (Unq^{104}), also called rutherfordium (Rf^{104}), and proceeds to the yet to be discovered element Uuo^{118}.

THE TRANSURANIC ELEMENTS

Some transuranic elements exist in nature, but most are artificially produced. They are radioactive and most of them are **transmutated** from one radioactive element into another element by the natural loss or gain of nuclear particles. Or, they are **synthesized** (man-made) by bombarding, or "slamming" together, particles or the nuclei of atoms of different elements that have two different atomic numbers and weights. These reactions require very high energies and are accomplished in **cyclotrons**, **nuclear reactors**, or high energy **particle accelerators**, which create a few atoms at a time of new, heavier, and very unstable elements. The transuranic elements are all radioactive. Some are found in small amounts in nature, but they are also synthesized by man. Those beyond the different isotopes of uranium were difficult to identify because of the number and large variety of fission products produced by the fission of uranium. Plutonium (Pu^{94}) is a product of the decay of uranium-238 and 239 into neptunium-239 and 238, which decays into plutonium-238 and can be produced as plutonium-239, which also spontaneously fissions. Plutonium-239 can be produced in a **breeder reactor** as both a source of energy or a nuclear fuel waste. Many scientists contributed to these discoveries and work that led to "atomic" energy and the "atomic" bomb, which should more accurately be called "nuclear" energy and the "nuclear" bomb, since the outer electron shells are not involved in the production of this energy as they are in chemical reactions.

THE TRANSACTINIDES

Since lawrencium was discovered and named (1961), a number of additional radioactive elements have been artificially produced. One might consider all the elements with atomic numbers higher than 103 as belonging to the actinide series. As mentioned, we divided the series into blocks. Therefore, we will consider the third block of elements beyond 103 as the **transactinides**. They have short half-lives, and because they are artificially produced in such small amounts, it is difficult to determine much about their properties and chemical characteristics. They range from atomic numbers 104 to 118. Interestingly, their structures were determined before they were discovered. It was predicted that the electrons in their $6d$ suborbital (of the 6th, or "P," shell) would increase from atomic numbers 104 to 118. The discovery of elements 110 (ununnilium) and 111 (unununilium) was announced in 1994. Number 112 (ununbiium) was discovered in 1996. Those with higher atomic numbers have not yet been discovered.

The listing of the newer and heaviest man-made elements will include most of the known information about them, but there is not as much known as there is for many of the lighter, well-known elements.

THE SUPERACTINIDES

Yet to be discovered are very heavy elements predicted by Glenn T. Seaborg (1912–   ) and his associates at the Lawrence-Livermore National Laboratory, located in Berkeley, California. They range in atomic number from 118 to 168. Dr. Seaborg refers to those in the range of 122 to 153 as **superactinides**. They could, theoretically, fit into the Periodic Table even though they are yet to be discovered or **synthesized**.

Block One
THE ACTINIDE ELEMENTS

ACTINIUM (Actinide Elements)
SYMBOL: Ac   **ATOMIC NUMBER:** 89   **PERIOD:** 7

COMMON VALENCE: 3 **ATOMIC WEIGHT:** 227 **NATURAL STATE:** Solid
COMMON ISOTOPES: There are no stable isotopes; all are radioactive. Actinium-227 emits **beta particles** and has a **half-life** of about 21.5 years. There are at least 11 other radioisotopes of actinium with half-lives from a few days to less than a minute. **PROPERTIES:** Actinium is the first element in the actinide series. The most stable of its isotopes, actinium-227, is a decay product from uranium-235. Actinium is about 10% more dense than lead. It is a whitish-silver metal with a melting point of 1050°C and a boiling point of 3200°C.

CHARACTERISTICS

Actinium is very closely related, both chemically and physically, to lanthanum, which is located just above it in the Periodic Table. Lanthanum is the first rare-Earth element of the lanthanide series.

ABUNDANCE AND SOURCE

Only a trace of actinium exists on the Earth as the result of the natural radioactive decay process of both uranium-235 and thorium-232.

Actinium is also produced by bombarding radium with neutrons in a **nuclear reactor**. It is found as minute traces in uranium ores.

Electron Configuration
K-2, L-8, M-18, N-32, O-18, P-9, Q-2

Actinium (Ac[89])

HISTORY

After the Curies had separated radium from uranium ore, their friend Andre-Louis Debierne (1874–1949) discovered another radioactive element mixed with uranium in the ore **pitchblende**. He named it after the Greek word *aktis*, which means "ray." However, in Latin, *actinium* means "radium."

COMMON USES

As a source of neutrons actinium-225 is used as a radioactive **tracer** in medicine and industry. It is too difficult to produce in substantial quantities to make it useful.

EXAMPLES OF COMMON COMPOUNDS

Most of the radioactive actinium isotopes that are produced in nuclear reactors are in milligram quantities. There are not many common compounds.

HAZARDS

Most of the radioactive isotopes of actinium pose an extreme radiation hazard. They are bone-seeking radioactive poisons.

❖ ❖ ❖

THORIUM (Actinide Elements)
SYMBOL: Th   **ATOMIC NUMBER:** 90   **PERIOD:** 7

COMMON VALENCE: 4 **ATOMIC WEIGHT:** 232.038 **NATURAL STATE:** Solid **COMMON ISOTOPES:** There are at least 12 known radioactive isotopes of thorium. None are stable. The most important radioactive isotope found in nature is thorium-232 with a **half-life** of over 14 billion years. **PROPERTIES:** Thorium is a dark-colored, soft, workable metal that can be rolled into sheets, **extruded**, and drawn into rods and wires. It is a reactive metal with a dark silver shine that is about as hard as lead. Density: approximately 12; melting point: 1700°C; boiling point: 4500°C.

CHARACTERISTICS
Thorium-232 has a half-life of 1.4×10^{10} (14 billion years).

Thorium can be converted to uranium-233 by bombarding it with neutrons.

Thorium is chemically similar to Hf and Zr, which are just above it in Group 4 of the Periodic Table. Thorium can form several compounds.

ABUNDANCE AND SOURCE
Thorium is the thirty-seventh most abundant element. It is found in monazite and thorite ores and is as abundant as lead. As a fuel for **nuclear reactors**, thorium has more energy potential than Earth's entire supply of uranium, coal, gas, and oil combined.

Electron Configuration
K-2, L-8, M-18, N-32, O-18, P-10, Q-2

Thorium (Th⁹⁰)

HISTORY
Jöns Jakob Berzelius (1779–1848), who discovered cerium in 1803, also discovered thorium in 1828. He named it *thorium* after Thor, the Norse mythological god of thunder.

COMMON USES
Thorium's most common use is as a "jacket" around the core of a nuclear reactor where it becomes **fissionable** uranium-233, which can then be used

for the nuclear reaction necessary to produce energy. It is also used in **photoelectric cells**, for X-ray tubes, and as an alloy with tungsten to make **filaments** for light bulbs.

Thorium may become more important as techniques are developed to convert it into uranium-233, which can then be used as a fissionable fuel in nuclear reactors. It has great potential to supplant all other nonrenewable energy sources.

EXAMPLES OF COMMON COMPOUNDS

Thorium carbide (ThC_2)—Used as a nuclear fuel.

Thorium chloride ($ThCl_4$)—Used to make **incandescent** lights.

Thorium dioxide (ThO_2)—Used in medicine and optical glass.

HAZARDS

As thorium undergoes natural radioactive decay, a number of products, including gases, are emitted. These decay products are extremely dangerous radioactive poisons if inhaled or ingested.

PROTACTINIUM (Actinide Elements)
SYMBOL: Pa   ATOMIC NUMBER: 91   PERIOD: 7

COMMON VALENCE: 4 and 5 **ATOMIC WEIGHT:** 231.036 **NATURAL STATE:** Solid **COMMON ISOTOPES:** There are no stable isotopes. There are about 13 or 14 unstable radioactive isotopes of protactinium ranging from protactinium-215 to protactinium-238. Only 2 occur naturally. They are radioactive protactinium-231 and protactinium-232. **PROPERTIES:** Protactinium is the third member of the actinide series. It is a silvery, **malleable** metal that does not **oxidize** in air. Protactinium melts at about 1600°C and boils at 4200°C. It has a density of 15.4.

CHARACTERISTICS

Of all its isotopes, protactinium-231 is the most important and has the longest **half-life** of 32,500 years.

Protactinium is an alpha (He^{++}) emitter. Chemically it resembles tantalum and niobium above it in the Periodic Table.

ABUNDANCE AND SOURCE

Protactinium exists in very small traces in the Earth's crust. It constitutes a very small amount of all uranium. Only about 350 milligrams can be extracted from each ton of high-grade uranium. It is the product of **fission** decay of uranium.

Protactinium can also be produced by the irradiation of thorium-230 in nuclear reactors or particle accelerators, which adds 1 proton and 1 or more neutrons to each thorium atom, thus changing element 90 into element 91.

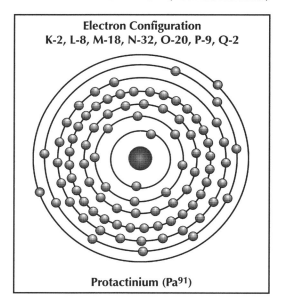

Electron Configuration
K-2, L-8, M-18, N-32, O-20, P-9, Q-2

Protactinium (Pa⁹¹)

HISTORY

Otto Hahn (1879–1968) and Lise Meitner (1878–1968) discovered a new radioactive element that decayed from uranium into element number 89, actinium. It was first named *brevium* for its short half-life. It was later named *protactinium* (element 91), which meant "before actinium."

This is another case where more than one or two people may have discovered the same element. Some references give credit to Frederick Soddy (1877–1956) and J. A. Carnston (birth date unknown), who discovered protactinium independently. Also, some authors give credit to K. Kajans and O. Gohring for the discovery of protactinium in 1913.

COMMON USES

Protactinium is very rare and not enough of it is available for commercial use. It is used only in laboratory research.

EXAMPLES OF COMMON COMPOUNDS

Protactinium forms many different compounds with the halogens, oxygen, carbon, and sulfur, and will form alloys. None of these are of commercial significance.

HAZARDS

All the isotopes of protactinium are highly radioactive and therefore very dangerous. It is a highly radioactive poison.

URANIUM (Actinide Elements)
SYMBOL: U   **ATOMIC NUMBER:** 92   **PERIOD:** 7

COMMON VALENCE: 3, 4, and 6 **ATOMIC WEIGHT:** 238.029 **NATURAL STATE:** Solid **COMMON ISOTOPES:** Isotopes of uranium range from uranium-222 to uranium-242. But there are only 3 natural unstable radioactive isotopes. They are uranium-234 with a **half-life** of 2.48×10^5 years; uranium-235 with a half-life of 7.13×10^8 years; and uranium-238 with a half-life of 4.51×10^9 years. U-238 makes up about 99% of all the uranium found in the Earth's crust. **PROPERTIES:** Uranium is a hard, dense, but workable (can be formed into different shapes), silver-colored metal that does not conduct electricity very well. It is a reactive metal that can form many compounds. Density: 19.0; melting point: 1132°C; boiling point: 3818°C.

CHARACTERISTICS

Uranium reacts with most nonmetallic elements and dissolves in some acids, but not alkalies.

Uranium is unique in that it can form solid solutions with other metals, such as molybdenum, titanium, zirconium, and niobium.

ABUNDANCE AND SOURCE

Uranium is the forty-eighth most abundant element on Earth. It is found in the ore **pitchblende**, which is uranium oxide (UO_2). It is also found in the ores of uraninite, coffinite, and carnotite, in Africa, France, Australia, and Canada, and in Colorado and New Mexico in the United States. It is found as a mixture of three different isotopes of uranium in nature as well as in compounds.

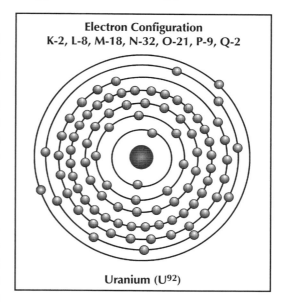

Electron Configuration
K-2, L-8, M-18, N-32, O-21, P-9, Q-2

Uranium (U^{92})

HISTORY

In February 1896 Antoine-Henri Becquerel (1852–1908) placed a piece of potassium uranyl sulfate on a photographic plate that was wrapped in black paper. When the plate was developed, it was fogged. This proved to be a newly discovered source of radiation other than sunlight or **fluorescence**.

In 1898 Marie Sklodowska Curie (1867–1934), who was interested in radiation, experimented with thorium and uranium. She coined the word "radioactivity" to describe this newly discovered type of radiation and went on to discover polonium and radium. Madame Curie and her husband, Pierre Curie (1859–1906), who discovered the **piezoelectric** effect which was used to measure radiation, and Henri Becquerel jointly received the Nobel Prize in physics.

COMMON USES

The most common use is to convert the rare uranium-235 isotope, which is the natural fissionable isotope of uranium, into plutonium through neutron capture. Plutonium, through controlled **fission**, is used in **nuclear reactors** to produce energy, heat, and electricity. Breeder reactors convert the more abundant, but nonfissionable, uranium-238 into the more useful, and fissionable, plutonium-239, which can be used both for the generation of electricity in nuclear power plants and to make nuclear weapons.

Although uranium forms compounds with many nonmetallic elements, there is not much use made of uranium outside of the nuclear energy industry.

EXAMPLES OF COMMON COMPOUNDS

Uranium-233—A fissionable isotope that is artificially produced by neutron capture of thorium-232. It is produced artificially and is an experimental fuel for nuclear breeder reactors.

Uranium-234—A natural isotope of uranium that is used in nuclear research.

Uranium-235—The rare, and the only naturally fissionable, isotope of uranium. It was used in the original nuclear (atomic) bombs. It takes a minimum of only 33 pounds of U-235 to reach a **critical mass**. The critical mass is the amount of fissionable material required to sustain a nuclear chain reaction, which will then become an "atomic" explosion. This 33-pound mass is a sphere about the size of a large softball. And, it takes only about 10 pounds of fissionable plutonium-239 to form a critical mass.

HAZARDS

The main hazard from all radioactive isotopes of uranium is radiation poisoning. Of course another potential hazard is using the isotopes of uranium to produce fissionable materials for other than peaceful purposes.

Block Two
THE TRANSURANIC ELEMENTS

NEPTUNIUM (Transuranic Elements)
SYMBOL: Np   **ATOMIC NUMBER:** 93   **PERIOD:** 7

COMMON VALENCE: 3, 4, 5, & 6 **ATOMIC WEIGHT:** 237.0482
NATURAL STATE: Solid **COMMON ISOTOPES:** There are 11 isotopes of

neptunium found in nature; all are radioactive. Neptunium-237 is an alpha emitter with a **half-life** of 2 million years. **PROPERTIES:** Neptunium is the first of the **transuranic elements**, i.e., those heavy, synthetic (man-made), radioactive elements following uranium in the actinide series of the Periodic Table. It is a silvery-white, reactive metal. Density: 20.45; melting point: 700°C; boiling point: 3900°C.

CHARACTERISTICS

The chemistry of neptunium is somewhat similar to that of both uranium and plutonium.

Neptunium can form compounds with most of the nonmetals, such as C, O, S, and the halogens.

ABUNDANCE AND SOURCE

Neptunium occurs as a minute trace in the Earth's crust. Naturally radioactive neptunium is found in very small amounts in uranium ores. It is also artificially produced in small amounts as a by-product during the commercial production of plutonium.

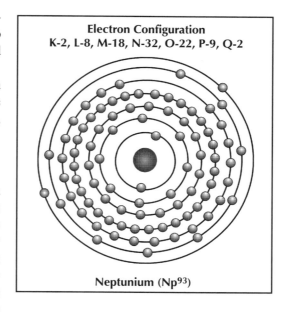

Electron Configuration
K-2, L-8, M-18, N-32, O-22, P-9, Q-2

Neptunium (Np⁹³)

Neptunium can be produced by bombarding uranium-238 with neutrons to produce neptunium-239.

HISTORY

In 1940 Edwin Mattison McMillan (1907–1991) and Philip Hauge Abelson (1913–1995) detected traces of element 93 while bombarding uranium with neutrons. The creation of artificially produced neptunium by bombarding uranium with neutrons could not be verified at the physics laboratory of the University of California, Berkeley. They named element 93 *neptunium* after the planet Neptune. Their 1940 experiments were continued by Arthur C. Wahl and Joseph W. Kennedy. Their work resulted in the reaction that produced detectable neptunium, as follows:

U-238 + H-2 (deuterons) → Np-239 + 2 neutrons; Np-238 by ß decay → Pu-238 (a radioisotope of element 94)

Of some interest is that uranium^{92} was named after the planet Uranus, and neptunium^{93}, which was discovered next, was named after the next planet in our solar system, Neptune. And finally, plutonium^{94}, the next transuranic element discovered, was named after Pluto, the next planet in our solar system.

COMMON USES
Neptunium is used in nuclear research and for instruments designed to detect neutrons. It can be converted into plutonium-238, which can be used in nuclear power plants.

EXAMPLES OF COMMON COMPOUNDS

Neptunium dioxide (NpO_2)—A powder is used to form metals for targets that are to be radiated by plutonium.

HAZARDS
All isotopes of neptunium are hazardous as radioactive poisons.

❖ ❖ ❖

PLUTONIUM (Transuranic Elements)
SYMBOL: Pu   **ATOMIC NUMBER:** 94   **PERIOD:** 7

COMMON VALENCE: 3, 4, 5, and 6   **ATOMIC WEIGHT:** 239.11   **NATURAL STATE:** Solid   **COMMON ISOTOPES:** The radioactive isotopes of plutonium range from plutonium-230 to plutonium-246. The most important, plutonium-239, has a **half-life** of 24,130 years.   **PROPERTIES:** Plutonium, through **fission**, produces both slow and fast neutrons that can be used to produce either electricity or nuclear bombs.

The **critical mass** of pure plutonium-239 is only 10 pounds. However, a mixture of appropriate materials that contain as little as 7% plutonium-239 can, without much difficulty, be used to make nuclear weapons that would fit into a small suitcase.

CHARACTERISTICS
Plutonium is a very reactive silvery metal.

Pu is the second **transuranic** element in the actinide series.

It **oxidizes** in air and is extremely poisonous.

ABUNDANCE AND SOURCE
Plutonium exists in trace amounts in nature. Most of the isotopes are man-made or produced by the natural decay of uranium.

Plutonium-239 is produced in **nuclear reactors** by bombarding uranium-238 with neutrons. The **transmutation** process follows:

U-238 + neutron → U-239 → decays to → Np-239 → decays to → Pu-239

HISTORY

Glenn Theodore Seaborg (1912–   ) recognized the transmutation of neptunium by its emission of **beta particles**. He realized that neptunium[93] was gaining a proton (atomic number) to become plutonium[94], which he named after the planet Pluto.

Plutonium is one of several new elements made by Seaborg's group of scientists at the University of California's **cyclotron**. They produced plutonium by bombarding uranium **oxide** with **deuterons** (nuclei of heavy hydrogen), which first produced neptunium that then decayed into plutonium. (See element number 93, neptunium, for the reaction that produced plutonium.)

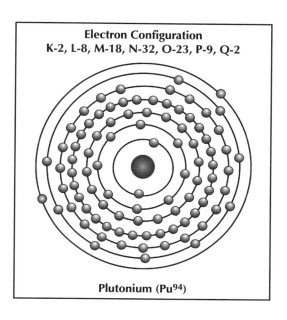

Electron Configuration
K-2, L-8, M-18, N-32, O-23, P-9, Q-2

Plutonium (Pu[94])

COMMON USES

The most common use of plutonium is as a "fuel" in nuclear reactors to produce electricity or for the critical mass required for sustaining **chain reactions** to produce nuclear weapons. Plutonium is used to convert non-fissionable uranium-238 into the isotope capable of sustaining a nuclear chain reaction.

It takes only 10 pounds of plutonium-239 to reach a critical mass to sustain fission and cause a nuclear explosion (compared to 33 pounds of uranium-235).

EXAMPLES OF COMMON COMPOUNDS

Plutonium can form compounds with nonmetals such as oxygen, the halogens, and nitrogen. It also forms many alloys.

Plutonium-239—The most important isotope because of its ability to fission in nuclear power reactors, as well as nuclear weapons.

HAZARDS

Plutonium is by far the most toxic radioactive poison known. The metal, its alloys, and its compounds must be handled in a shielded and enclosed "glove box" that contains an inert argon atmosphere. It is a **carcinogen** that can cause radiation poisoning leading to death.

AMERICIUM (Transuranic Elements)
SYMBOL: Am   **ATOMIC NUMBER:** 95   **PERIOD:** 7

COMMON VALENCE: 3 (5 and 6) **ATOMIC WEIGHT:** 241 **NATURAL STATE:** Solid **COMMON ISOTOPES:** The most important isotope is americium-241, an alpha emitter, with a **half-life** of 450 years. There are 14 other radioactive isotopes ranging from americium-232 to americium-247, with half-lives ranging from about 1 hour to over 7,000 years. **PROPERTIES:** Americium is a synthetic radioactive **transuranic** element of the actinide series that has similar characteristics to the rare-Earths. It is a **malleable** metal with properties similar to those of lead. Density: 13.7; melting point: 1000°C; boiling point: 2600°C.

CHARACTERISTICS

Although the common valence of americium-241 is 3, americium is **tetravalent**, i.e., has several other valences when oxidized.

ABUNDANCE AND SOURCE

Americium does not exist in nature. It is all manmade. Radioisotope americium-241 is separated from used plutonium after the "spent" plutonium is removed from **nuclear reactors**.

Plutonium, through **beta decay**, produces both radioactive americium-241 and americium-243 within nuclear reactors.

HISTORY

In 1944 Glenn T. Seaborg (1912–   ) and his associ-

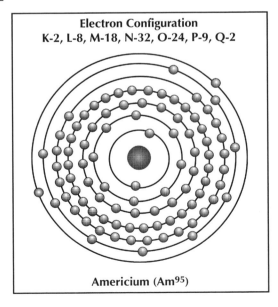

Electron Configuration
K-2, L-8, M-18, N-32, O-24, P-9, Q-2

Americium (Am^{95})

ates, R. A. James, L. O. Morgan, and A. Ghiorso, prepared americium from plutonium by bombarding plutonium with **alpha particles** (helium nuclei) and neutrons. This converted Pu^{94} into the isotope Am^{95}.

Since these experiments were performed in the **cyclotron** at the University of California at Berkeley, they named it *americium* after their country, America.

COMMON USES

Americium is used as a **gamma ray** source for **radiography**. It is used in electronic devices, and as a gamma ray source it is used as a diagnostic aid to check the quality of welds in metals.

It is also used as a neutron target in instruments that can measure the level of fluids in tanks.

EXAMPLES OF COMMON COMPOUNDS

Americium-243—Long-lived (half-life of 7,400 years). Used to prepare curium-244.

Americium can form compounds with oxygen, lithium, hydrogen, and the halogens, as well as with several other nonmetals.

HAZARDS

All of the radioisotopes of americium are deadly sources of radiation and cause radiation poisoning. Precautions must be taken when working with it.

❖   ❖   ❖

CURIUM (Transuranic Elements)
SYMBOL: Cm   **ATOMIC NUMBER:** 96   **PERIOD:** 7

COMMON VALENCE: 3 and 4   **ATOMIC WEIGHT:** 244   **NATURAL STATE:** Solid   **COMMON ISOTOPES:** All 16 of the radioactive isotopes of curium are man-made. They range from curium-236 to curium-251. **PROPERTIES:** Curium is a silvery-white heavy metal that is chemically more reactive than aluminum and has properties similar to those of uranium and plutonium. Density: 13.5; melting point: 340°C; boiling point: 3500°C.

CHARACTERISTICS

Curium is a synthetic, artificial **transuranic** element of the actinide series that has some characteristics of the rare-Earth elements in the lanthanide series.

ABUNDANCE AND SOURCE

There is no natural curium on the Earth. All of its isotopes are man-made artificially. Curium metal can be produced by the reduction of curium trifluoride with barium vapor.

Curium is also produced by bombarding plutonium-239 with **alpha particles** (nuclei of helium atoms).

HISTORY

At the same time in 1944 that Glenn T. Seaborg (1912–   ) and his colleagues were producing the new element americium with the cyclotron at Berkeley, California, they created curium. By shooting **subatomic particles** at Pu^{94}, they created Cm^{96}. This new element was named *curium*, after the Curies, who are considered pioneers of radioactivity research.

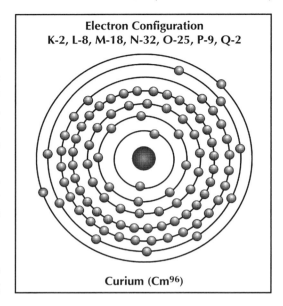

Electron Configuration
K-2, L-8, M-18, N-32, O-25, P-9, Q-2

Curium (Cm^{96})

COMMON USES

The most important use of curium-244 is to provide the power for small, compact **thermoelectric** sources of electricity, by generating heat through the nuclear decay of the radioisotope. These small, efficient power sources can be used in remote locations on the Earth to provide electricity to areas that cannot secure it from other sources, and as a source of electricity in spacecraft.

EXAMPLES OF COMMON COMPOUNDS

Curium can form compounds with oxygen, the halogens, carbon, and several other nonmetals. They are very scarce.

HAZARDS

Curium isotopes are radioactive bone-seeking poisons that attack the skeletal system of humans and animals.

❖　❖　❖

BERKELIUM (Transuranic Elements)
SYMBOL: Bk   **ATOMIC NUMBER:** 97   **PERIOD:** 7

COMMON VALENCE: 2 and 3   **ATOMIC WEIGHT:** 249   **NATURAL STATE:** Solid   **COMMON ISOTOPES:** Berkelium is an artificially made radioactive element. There are 10 radioisotopes, ranging from berkelium-242 to berkelium-251. **PROPERTIES:** Berkelium is a **transuranic** element of the actinide series. As a metal, it is chemically reactive and is very similar to curium-244. Density: 14.8; melting point: about 998°C; boiling point: unknown.

CHARACTERISTICS

In its **oxidation** state, berkelium is chemically similar to the rare-Earth terbium, which is located just above it in the lanthanide series.

ABUNDANCE AND SOURCE

Berkelium does not exist naturally on the Earth. It is produced as radioactive isotopes in **cyclotrons** (atom smashers) and **nuclear reactors**.

HISTORY

In 1949, Glenn T. Seaborg (1912–   ) and his colleagues, S. G. Thomson

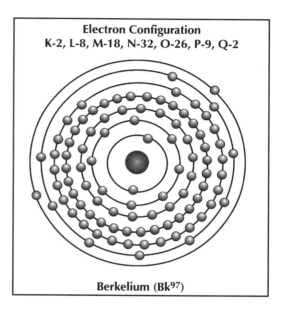

Electron Configuration
K-2, L-8, M-18, N-32, O-26, P-9, Q-2

Berkelium (Bk^{97})

and A. Ghiorso, continued to use the cyclotron at Berkeley, California, to create new transuranic radioactive elements. They bombarded americium^{95} with alpha particles (helium nuclei or He^{++} ions) to form a new element (97). Since this new element was so closely related to terbium (Tb^{65}), which was named for the town of Ytterby in Sweden, its source, they decided to name their new element *berkelium* after their town, Berkeley, in California.

COMMON USES

There is not much use for berkelium, except as a source to produce a radioisotope of californium. Other uses are for laboratory experimentation.

EXAMPLES OF COMMON COMPOUNDS

Berkelium oxide (BkO$_2$), berkelium trifluoride (BkF$_3$), and a few other compounds have been formed, but they are difficult to produce.

HAZARDS

Like other radioactive isotopes, berkelium metal and its compounds are extremely dangerous. Because they are sources of strong radiation, precautions must be taken when working with them.

❖ ❖ ❖

CALIFORNIUM (Transuranic Elements)
SYMBOL: Cf   **ATOMIC NUMBER:** 98   **PERIOD:** 7

COMMON VALENCE: 3 **ATOMIC WEIGHT:** 252 **NATURAL STATE:** Solid **COMMON ISOTOPES:** There are no natural isotopes of californium. The artificial radioisotopes range from californium-239 to californium-255. The 2 most common isotopes are californium-252 and californium-249. Californium-252 with a **half-life** of 2.6 years is the easiest to produce in a **nuclear reactor**. **PROPERTIES:** Californium is a synthetic radioactive transuranic element of the actinide series. The pure metal form has not yet been produced. All of its compounds and isotopes are scarce.

CHARACTERISTICS

All of the isotopes of californium are the result of artificial **transmutation** of radioactive isotopes of several higher elements.

ABUNDANCE AND SOURCE

There are no natural forms of californium in the Earth's crust. The artificial isotopes and their compounds are produced in only very small quantities.

Californium is the result of transmutation of higher elements.

HISTORY

In 1949, Glenn T. Seaborg (1912–) and his

Electron Configuration
K-2, L-8, M-18, N-32, O-27, P-9, Q-2

Californium (Cf 98)

associates, who discovered several new elements by using the Berkeley cyclotron, also produced element 98. This was accomplished by bombarding curium-242 with alpha particles, which produced californium-252.

Since they did their work in California, they called this new element *californium*.

COMMON USES

Californium's uses are rather limited. It is used to search for and test minerals, for logging (locating and marking) oil wells, and in medicine.

Californium spontaneously fissions, which makes it an ideal and accurate counter for electronic systems.

EXAMPLES OF COMMON COMPOUNDS

A few compounds are known. They are oxides, chlorides, and fluorides. They have little practical use.

HAZARDS

Californium's greatest danger is as a biological bone-seeking element, which can be both a radiation hazard and a useful treatment for bone cancer. Radiation poisoning is a potential hazard.

EINSTEINIUM (Transuranic Elements)
SYMBOL: Es   **ATOMIC NUMBER:** 99   **PERIOD:** 7

COMMON VALENCE: 2   **ATOMIC WEIGHT:** 253   **NATURAL STATE:** Solid
COMMON ISOTOPES: There are no natural forms of einsteinium. The artificial radioisotopes of einsteinium range from einsteinium-243 to einsteinium-256. The **half-lives** of its radioisotopes range from a few seconds to about one year. The most stable isotope is einsteinium-254.   **PROPERTIES:** Einsteinium is the heaviest transuranic element in the actinide series. The metal form is so rare that there is barely enough for it to be weighed. Melting point: 850°C.

CHARACTERISTICS

Einsteinium is very reactive and volatile.

It has chemical properties similar to those of the rare-Earth holmium, located above it in the lanthanide series.

ABUNDANCE AND SOURCE

No einsteinium exists in nature. It is all produced by artificial nuclear **transmutation** by higher radioactive elements.

Einsteinium is artificially produced in **cyclotrons** by bombarding uranium with high-speed nuclei of nitrogen. It is also produced in nuclear reactors by irradiation of Cf and Pu with neutrons.

HISTORY

Einsteinium was first discovered in 1952 in the material left over from the first hydrogen (thermonuclear) bomb. This bomb was a **fusion** (combining) reaction of heavy hydrogen where the deuterium (heavy hydrogen) nuclei were driven together by great forces, as compared to **fission** (splitting) of heavy nuclei, as in the atomic bomb. This was the first time the more complex elements 99 and 100 were formed. Einsteinium was named after Albert Einstein to honor him for developing the concept that energy and matter are essentially the same, represented by his famous formula $E = mc^2$, which, in theory, made the atom bomb possible.

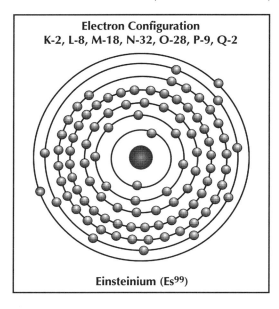

Electron Configuration
K-2, L-8, M-18, N-32, O-28, P-9, Q-2

Einsteinium (Es^{99})

COMMON USES

Einsteinium does not really have any common uses. It is used in radioactive chemistry research.

EXAMPLES OF COMMON COMPOUNDS

Not many compounds are formed or known.

HAZARDS

The radioisotopes of einsteinium are highly unstable and the small amount of the metal available is extremely reactive.

FERMIUM (Transuranic Elements)
SYMBOL: Fm   **ATOMIC NUMBER:** 100   **PERIOD:** 7

COMMON VALENCE: 3   **ATOMIC WEIGHT:** 254   **NATURAL STATE:** Solid
COMMON ISOTOPES: There is no natural fermium on the Earth. The artificial, man-made isotopes are all radioactive. They range from fermium-242 to fermium-259. Their **half-lives** range from only milliseconds to 100 days. Fermium-257 is the most stable of all the radioactive isotopes of fermium.

PROPERTIES: Less than one-millionth of a gram of fermium has been artificially produced. Fermium-244, fermium-256, and fermium-258 all decay by spontaneous **fission** (where an isotope of one element "splits" to become several different elements with different atomic numbers and/or weights).

CHARACTERISTICS

Fermium's chemical properties are not very well known, but they are similar to those of the rare-Earth erbium, which is just above it in the lanthanide series.

ABUNDANCE AND SOURCE

Fermium, which is not found in nature, is produced by bombarding Pu or Es with neutrons in a **cyclotron**. Fm can also be artificially produced in **nuclear reactors** that "shoot" oxygen ions (positively charged nuclei of oxygen) at uranium.

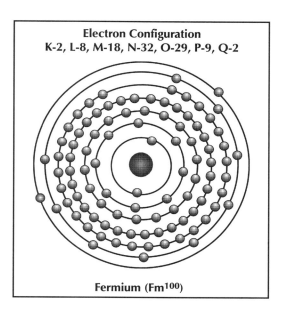

Electron Configuration
K-2, L-8, M-18, N-32, O-29, P-9, Q-2

Fermium (Fm^{100})

HISTORY

Like einsteinium, fermium was found in debris left over from the first thermonuclear **fusion** bomb set off in 1952. Fusion is somewhat the opposite of fission. While fission splits the nuclei of one element into nuclei of smaller elements, fusion combines the nuclei of heavy hydrogen into the nuclei of the element helium and other elements. Both reactions produce much energy, radiation, and many different subatomic particles.

Fermium was named for Enrico Fermi (1901–1954), who discovered artificial radioactivity by shooting neutrons into elements. Following is an interesting reaction that produces Lr, Md, and Fm transuranic elements by particle radiation:

Am-243(95) + O-18(8) + 5 neutrons → Lr-256(103) which decays
by α loss → Md-252(101) + electron capture → Fm-252$^{(100)}$

COMMON USES

Fermium is used for **tracer** studies and laboratory research.

EXAMPLES OF COMMON COMPOUNDS
None are known.

HAZARDS
All the radioisotopes of fermium pose radiation hazards. There is not much chance of coming in contact with one of fermium's isotopes since they all have very short half-lives of only a few seconds, and thus don't exist long enough to be a hazard.

❖ ❖ ❖

MENDELEVIUM (Transuranic Elements)
SYMBOL: Md   **ATOMIC NUMBER:** 101   **PERIOD:** 7

COMMON VALENCE: 3 **ATOMIC WEIGHT:** 256 **NATURAL STATE:** Solid **COMMON ISOTOPES:** The known isotopes of mendelevium range from mendelevium-248 to mendelevium-258. Their half-lives range from a few seconds to about 60 days for the heaviest, mendelevium-258. **PROPERTIES:** Mendelevium is a synthetic radioactive metallic element that is artificially produced by nuclear **transmutation**, which is accomplished by bombarding other isotopes with charged particles. Not enough has been produced to know its properties.

CHARACTERISTICS
Mendelevium has chemical properties similar to those of thulium, which is located just above it in the lanthanide series.

Mendelevium decays through natural spontaneous **fission** to form other isotopes.

ABUNDANCE AND SOURCE
Only trace amounts of mendelevium have been artificially produced. Only several million atoms have been made, which is not enough to weigh, so other laboratory methods are used to study it, such as ion-exchange **chromatography** and **spectroscopy**.

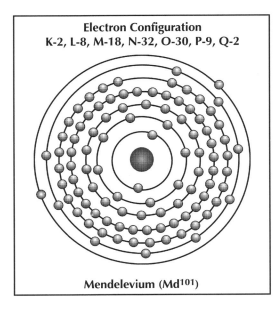

Electron Configuration
K-2, L-8, M-18, N-32, O-30, P-9, Q-2

Mendelevium (Md[101])

HISTORY

In 1955 Glenn T. Seaborg (1912–), A. Ghiorso, and his other associates bombarded their newly artificially made element einsteinium-253 with high-speed positive protons and formed a few atoms, one at a time, of another new element with atomic number 101, i.e., mendelevium-256.

They called this new element *mendelevium* after D. I. Mendeleyev (1834–1907), who is considered the father of the Periodic Table. He established the periodic nature of elements and arranged them in "periods" according to the ascending order of their atomic weights (mass) and "grouped" them according to their similar properties.

COMMON USES

Not much is known about mendelevium, and it has no uses other than for laboratory research.

EXAMPLES OF COMMON COMPOUNDS

None are known.

HAZARDS

Like any artificially produced radioactive isotope that goes through natural fission, mendelevium is an extreme radiation hazard. But little of it exists, so there is not much danger of individual or public radiation poisoning.

❖   ❖   ❖

NOBELIUM (Transuranic Elements)

SYMBOL: No   **ATOMIC NUMBER:** 102   **PERIOD:** 7

COMMON VALENCE: 3   **ATOMIC WEIGHT:** 254   **NATURAL STATE:** Solid
COMMON ISOTOPES: There are 9 artificially produced radioactive isotopes of nobelium. They range from nobelium-250 to nobelium-262. The most common is nobelium-255. Their **half-lives** are so short that not much is known about their properties. **PROPERTIES:** Only a few atoms of nobelium have been artificially produced in **cyclotrons** and its properties are not well known.

CHARACTERISTICS

Nobelium's chemical characteristics are not known, but it seems reasonable to assume that they resemble the properties of ytterbium[70], which is located just above it in the lanthanide series.

ABUNDANCE AND SOURCE

Nobelium does not exist in nature. All of its isotopes are radioactive, unstable, and man-made.

Nobelium is the tenth transuranic element in the actinide series. It is produced in **cyclotrons** by bombarding curium with carbon-13 nuclei that have been accelerated to high speeds.

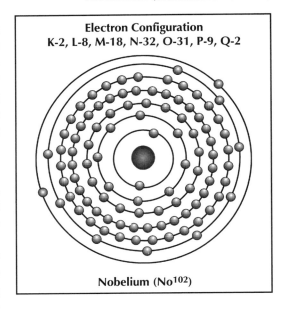

Electron Configuration
K-2, L-8, M-18, N-32, O-31, P-9, Q-2

Nobelium (No[102])

HISTORY

Researchers in Russia, Sweden, and California have all made claims of discovering nobelium.

Credit for the discovery of nobelium is usually given to Glenn T. Seaborg (1912–)and his colleagues at the Lawrence Radiation Laboratory in Berkeley, California, who, in 1958, artificially produced it by using carbon nuclei to bombard curium, which produced the radioisotope nobelium-254. Even so, internationally, credit for No's discovery is shared with the Dubna group in Russia. In addition, a team of scientists at the Nobel Institute of Physics in Sweden claimed they discovered nobelium in 1957. Their claim was not independently confirmed, but they selected the name, nobelium, which the Berkeley group agreed to keep. The Commission on Nomenclature of Inorganic Chemistry of the IUPAC (International Union of Pure and Applied Chemistry) gave it the name *nobelium* after Alfred B. Nobel, who developed and manufactured explosives. He also established the Nobel Prizes for excellence in a variety of fields, including the sciences.

COMMON USES

None are known.

EXAMPLES OF COMMON COMPOUNDS

None are known.

HAZARDS

Although nobelium poses a radiation hazard, the chances of being exposed to it are nil since there is so little of it and its half-life is only a few seconds.

❖   ❖   ❖

LAWRENCIUM (Transuranic Elements)
SYMBOL: Lr or Lw   ATOMIC NUMBER: 103   PERIOD: 7

COMMON VALENCE: 3 (?) **ATOMIC WEIGHT:** 257 **NATURAL STATE:**
Solid **COMMON ISOTOPES:** The artificially produced radioactive isotopes
of lawrencium range from lawrencium-253 to lawrencium-262. Their
half-lives range from lawrencium-257 at 8 seconds to lawrencium-260 at 3
minutes. **PROPERTIES:** Lawrencium's isotopes radiate **alpha particles** as
they decay spontaneously. The very small amount of lawrencium that is
artificially produced does not exist long enough to learn more about its
chemical and physical
properties.

CHARACTERISTICS
   Lawrencium is the last of
the transuranic elements and
the fifteenth in the actinide series, just as there are 15 elements in the lanthanide series.

**ABUNDANCE
AND SOURCE**
   Lawrencium does not exist
in nature. It is all artificially
produced by bombarding californium atoms with the positively charged nuclei of boron.
Very small amounts have been
produced.

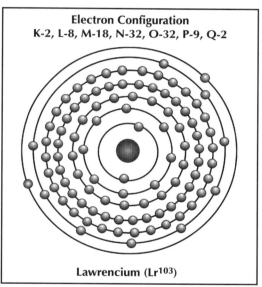

Electron Configuration
K-2, L-8, M-18, N-32, O-32, P-9, Q-2

Lawrencium (Lr^{103})

HISTORY
   In 1961 just a few atoms of lawrencium were produced in the **cyclotron** at the Lawrence Radiation Laboratory in Berkeley, California, by
A. Ghiorso, T. Sikkeland, A. E. Larsh and R. M. Latimer. The reaction that
produced lawrencium follows:

   Cf-249 & 252 + B-10 & 11 + several neutrons → Lr-257 & 258
   (element 103) + α radiation.

The nuclei of Lr-257, 258 have half-lives of less than a few seconds. The
old (original) symbol for lawrencium was Lw, but it was changed to Lr by
the naming committee of the IUPAC.

Lawrencium was named for Ernest Orlando Lawrence (1901–1958), who invented the cyclotron and for whom the famous laboratory located on the campus of the University of California, Berkeley, was named.

COMMON USES
None are known.

EXAMPLES OF COMMON COMPOUNDS
None are known.

HAZARDS
Like the other short-lived radioactive isotopes, lawrencium is a radiation hazard. Also, as with the others, the danger to individuals and the public is small since there is not much of it in existence, and that small amount has a short half-life of a fraction of a second.

INTRODUCTION: THE TRANSACTINIDE ELEMENTS

A number of very unstable radioactive elements beyond the **transuranic** elements have been discovered. They are referred to as the **transactinides** (beyond the actinides) because they extend beyond lawrencium. They do not exist in nature and have very short half-lives. In some cases, only one or a few atoms have been man-made, and those exist for only a few minutes to just a fraction of a second. Thus, it is extremely difficult to learn much about their physical properties and chemical characteristics.

Darleane C. Hoffman, Director of the G. T. Seaborg Institute for Transactinium Science at the Lawrence Livermore National Laboratory in Berkeley, California, published an article titled, "The Heaviest Elements," which explains the difficulty of determining the chemistry of heavy elements. She writes, "How long does an atom need to 'exist' before it's possible to do any meaningful chemistry on it? Is it possible to learn anything at all about the reactions of an element for which no more than a few dozen atoms have ever existed simultaneously?" Dr. Hoffman and her students and other laboratories worldwide produce one atom at a time in accelerators by bombarding radioactive targets with high-intensity beams of heavy ions (charged atoms and particles of other elements). All of the heavy elements decay, and most have half-lives of a few minutes to only fractions of milliseconds. Due to these conditions, limited chemical studies have been performed on these short-lived radioactive elements, but new nuclear chemistry techniques are used to investigate the properties of these scarce elements, which spontaneously fission (SF) and thus have very short half-lives.

PERIODIC TABLE OF THE ELEMENTS

GROUPS →	1 IA	2 IIA	3 IIIB	4 IVB	5 VB	6 VIB	7 VIIB	8 VIII	9 VIII	10 VIII	11 IB	12 IIB	13 IIIA	14 IVA	15 VA	16 VIA	17 VIIA	18 VIIIA
1	1 H 1.0079																	2 He 4.00260
2	3 Li 6.941	4 Be 9.01218											5 B 10.81	6 C 12.011	7 N 14.0067	8 O 15.9994	9 F 18.9984	10 Ne 20.179
3	11 Na 22.9898	12 Mg 24.305											13 Al 26.9815	14 Si 28.0855	15 P 30.9738	16 S 32.066(6)	17 Cl 35.453	18 Ar 39.948
4	19 K 39.0983	20 Ca 40.08	21 Sc 44.9559	22 Ti 47.88	23 V 50.9415	24 Cr 51.996	25 Mn 54.9380	26 Fe 55.847	27 Co 58.9332	28 Ni 58.69	29 Cu 63.546	30 Zn 65.39	31 Ga 69.72	32 Ge 72.59	33 As 74.9216	34 Se 78.96	35 Br 79.904	36 Kr 83.80
5	37 Rb 85.4678	38 Sr 87.62	39 Y 88.9059	40 Zr 91.224	41 Nb 92.9064	42 Mo 95.94	43 Tc (98)	44 Ru 101.07	45 Rh 102.906	46 Pd 106.42	47 Ag 107.868	48 Cd 112.41	49 In 114.82	50 Sn 118.71	51 Sb 121.75	52 Te 127.60	53 I 126.905	54 Xe 131.29
6	55 Cs 132.905	56 Ba 137.33	★	72 Hf 178.49	73 Ta 180.948	74 W 183.85	75 Re 186.207	76 Os 190.2	77 Ir 192.22	78 Pt 195.08	79 Au 196.967	80 Hg 200.59	81 Tl 204.383	82 Pb 207.2	83 Bi 208.980	84 Po (209)	85 At (210)	86 Rn (222)
7	87 Fr (223)	88 Ra 226.025	▲	104 Und (261)	105 Unp (262)	106 Unh (263)	107 Uns (264)	108 Uno (265)	109 Une (266)	110 Uun (267)	111 Uuu (272)	112 Uub	113 Uut	114 Uuq	115 Uup	116 Uuh	117 Uus	118 Uuo

← TRANSITION ELEMENTS

6 ★ Lanthanide Series (RARE EARTH)

57 La 138.906	58 Ce 140.12	59 Pr 140.908	60 Nd 144.24	61 Pm (145)	62 Sm 150.36	63 Eu 151.96	64 Gd 157.25	65 Tb 158.925	66 Dy 162.50	67 Ho 164.930	68 Er 167.26	69 Tm 168.934	70 Yb 173.04	71 Lu 174.967

7 ▲ Actinide Series (RARE EARTH)

89 Ac 227.028	90 Th 232.038	91 Pa 231.036	92 U 238.029	93 Np 237.048	94 Pu (244)	95 Am (243)	96 Cm (247)	97 Bk (247)	98 Cf (251)	99 Es (252)	100 Fm (257)	101 Md (258)	102 No (259)	103 Lr (260)

The naming of the heavy transactinide elements has caused some controversy. Different people in different laboratories in different countries have claimed discovery of the same elements, and thus the right to name them. This leads to different names for a few of the same elements. The naming of the new transfermium elements (beyond 100) was assigned to a working group by the International Union of Pure and Applied Chemistry (IUPAC). Some of the discoveries for the same element list slightly different atomic weights (isotope) and half-lives. The name for element 106 was proposed to honor Glenn T. Seaborg, who discovered plutonium and several other heavy elements. Later, element 106's name was changed to rutherfordium (Rf), which was the original name for element 104. This controversy and naming system becomes confusing. In addition to naming the transactinide elements after people, there is a common naming system devised by the IUPAC that is based on the symbol *un* (for one), which is repeated with other abbreviations included. For instance: 0=nil, 1=un, 2=bi, 3=tri, 4=quad, 5=pent, 6=hex, 7=sept, 8=oct, 9=enn. The *un* is repeated for the first two digits when elements reach 110 and beyond.

Now it gets interesting. Since IUPAC does not allow naming an element after a living person, they disallowed the name for 106, seaborgium (Sg), because Dr. Seaborg was still alive at that date. But since then, at least in the United States, 106 is still called seaborgium. In a 1994 meeting, the IUPAC created more confusion by renaming all the transactinide elements from 104 through 109. Their proposed new names follow:

Element 104: Rutherfordium, Rf

Element 105: Dubnium, Db

Element 106: Seaborgium, Sg

Element 107: Bohrium, Bh

Element 108: Hassium, Hs

Element 109: Meitnerium, Mt

Elements 110–112: unnamed

It should be mentioned that the American Chemical Society (ACS) rejected this list of names.

If the scientists are confused, think about the members of the general public who are interested in these matters. The problems of determining discovery and naming of heavy transactinide elements may be solved in the near future (the original date for a resolution to these problems was 1997).

When examining the following information of the latest discovered transactinide elements, it might be interesting to compare their properties with the elements located above them in the Periodic Table, called their **homologues**. According to theory, Unq^{104} should be somewhat similar to hafnium^{72}, and so forth. The heavy elements 113 through 118 exist only in theory. They have not been identified, but, no doubt, their discovery will be announced by Dr. Hoffman, her colleagues, and her students in the near future.

Block Three
THE TRANSACTINIDE ELEMENTS

UNNILQUADIUM: Rutherfordium (Transactinide Elements)

May be interpreted as *un-nil-quad-ium*, which stands for 1-0-4 + ium as a standard ending, which is the Greek suffix for chemical elements. (It is also known as rutherfordium Rf, the accepted name in the United States. It is known as kurchatovium Ku in Russia, and in 1994 the IUPAC proposed the name dubnium, which has not yet been accepted.)

SYMBOLS: Unq and Rf   **ATOMIC NUMBER:** 104

ATOMIC WEIGHT: Ranges from 253 to 263   **COMMON ISOTOPES:** Eleven different isotopes of element 104, unnilquadium (or Rf), have been identified. Not all have been confirmed. They are 253, 254 (?), 255, 256, 257, 258, 259, 260, 261, 262, and 263 (?).

PROPERTIES AND CHARACTERISTICS

Unnilquadium spontaneously fissions (SF), producing alpha radiation. Thus, its isotopes have half-lives that are reported as varying from 0.5 milliseconds to 65 seconds. Its chemical properties are related to hafnium, located just above it in the Periodic Table. It was first produced by the Joint Institute for Nuclear Research (JINR) located in Dubna, Russia, in a cyclotron reaction that collided Pu-242 with Ne-22 + 2 neutrons to produce Unq-260 (or Ku-260), plus some alpha radiation. This element's SF and confirmation were disputed by scientists at the Lawrence Radiation Laboratory in Berkeley, who used high energy particle accelerators to produce three different isotopes of element 104, unnilquadium (or Rf):

Cf-249 + C-12 + 4 neutrons = Unq-253 + α radiation
Cf-249 + C-13 + 3 neutrons = Unq-255 + α radiation
Cm-248 + O-16 + 5 neutrons = Unq-259 + α radiation

HISTORY

Unnilquadium, as a theoretical new element, was originally named by the International Union of Pure and Applied Chemistry (IUPAC), which is located in Oxford, England. Unq-104's discovery was claimed by the JINR in Dubna, Russia, in 1964. They named it *kurchatovium* (Ku), for the head of their center, Ivan Kurchatov. Others could not verify the Russian claim, due to the nature of the SF of this heavy transactinide element. The Berkeley group produced isotopes of 104 in 1969, which could be confirmed by experimental fission detection equipment the scientists developed. The Berkeley group developed an "automated rapid chemistry apparatus," the "heavy-element volatility instrument," and the "merry-go-round." This equipment serves as a sensitive particle/radiation display detection system that is used to identify and confirm the chemical and physical properties of the decay particles of heavy elements. The Berkeley group claimed that since they identified and confirmed new isotopes of Unq, it should be named after the New Zealand physicist/chemist Ernest Rutherford. The IUPAC agreed, and it was officially given the symbol Rf for *rutherfordium*—at least for the next few years. Element 104 is still called rutherfordium (Rf) by the Berkeley group and IUPAC. Ten different isotopes of Rf have been identified with half-lives ranging from 0.5 milliseconds to 65 seconds. Rf is considered an eka-hafnium element.

COMMON USES
None, except for high-energy nuclear/particle research.

EXAMPLES OF COMMON COMPOUNDS
None are published.

HAZARDS
The same as for any radioactive element. But since only a few atoms are produced at a time, for such a short time, there is no great danger.

UNNILPENTIUM: Dubnium (Transactinide Elements)
Interpreted as 1-0-5 + ium (the Greek suffix for chemical elements).
SYMBOLS: Unp and Db   **ATOMIC NUMBER:** 105

ATOMIC WEIGHT: Ranges from 255 to 263   **COMMON ISOTOPES:** There are 7 confirmed isotopes of element 105. They are 255, 257, 258, 260, 261, 262, and 263.

PROPERTIES AND CHARACTERISTICS
Unnilpentium spontaneously fissions (SF) and thus exists for only a short time, ranging from 1.3 seconds to 34 seconds. Alpha particles (radiation) are produced by the SF, while two of the isotopes (258 and 262) may be used for detection purposes. As with most heavy transactinide elements, individual atoms are produced only one, or just a few, at a time, and they

exist for such a short period of time that it is very difficult to determine their chemical and physical properties.

HISTORY

As with several other transactinide elements, the discovery, and thus naming of number 105, is in dispute. Before it was discovered, the IUPAC named it *unnilpentium* (Unp). In 1970 the Dubna, Russia, group reported the following reaction:

$$Am\text{-}243 + Ne\text{-}22 + 5 \text{ neutrons} = Unp\text{-}260 + \alpha \text{ radiation}$$

They named their element 105, *nielsbohrium* (Ns). The quality of their data was suspect. They made corrections in their SF data, but others could not correlate the Russian's results with the homologue of Nb and Ta. The homologues of transactinide elements are located just above them in the Periodic Table, and the transactinides are expected to exhibit similar chemistry and physics as their homologues.

In 1970 and 1971 the Berkeley group provided positive evidence for the production of isotopes of 105 in their Heavy Ion Linear Accelerator (HILAC). The production of three of the isotopes are exhibited in the following reactions:

$$Cf\text{-}249 + N\text{-}15 + 4 \text{ neutrons} = Unp\text{-}260 + \alpha \rightarrow Lr\text{-}256 + \alpha$$
$$Cf\text{-}250 + N\text{-}15 + 4 \text{ neutrons} = Unp\text{-}261 + \alpha \rightarrow Lr\text{-}257 + \alpha$$
$$Bk\text{-}249 + O\text{-}18 + 5 \text{ neutrons} = Unp\text{-}262 + \alpha \rightarrow Lr\text{-}258 + \alpha$$

Note: the $\rightarrow$ represents a further decay of element 105 (Ha) to element 103, Lawrencium (Lr), and α is an alpha particle (He++).

Due to the Berkeley evidence, the American Chemical Society (ACS) recommended that element 105 be named *hahnium* (Ha), after the German radiochemist Otto Hahn. A joint committee of IUPAC and the International Union of Pure and Applied Physics (IUPAP) met to consider conflicting claims for both elements 104 and 105. The committee, including three Russian members, recommended that scientists from the Russian and Berkeley groups get together to review and confirm conflicting data. This cooperative venture was not implemented. Subsequently, the Western world accepted the recommendation of the IUPAC to name element 105 *dubnium* (Db).

COMMON USES

None, except for research purposes.

EXAMPLES OF COMMON COMPOUNDS
None are published, but it could possibly form compounds similar to its homologue, element 73, tantalum.

HAZARDS
Similar to all highly radioactive isotopes, but no threat to the public.

❖ ❖ ❖

UNNILHEXIUM (Transactinide Elements)
May be interpreted as 1-0-6 + ium (for the Greek suffix for chemicals). In the United States it is also known as seaborgium. More recently a name change was proposed to seaborgium.

SYMBOLS: Unh and Sg   **ATOMIC NUMBER:** 106

ATOMIC WEIGHT: Ranges from 258 to 266.   **COMMON ISOTOPES:** There are 7 known isotopes of Unh. They are 258, 259, 260, 261, 263, 265 (?), and 266 (?).

PROPERTIES AND CHARACTERISTICS
Isotopes of unnilhexium's spontaneous fission (SF) range from 0.4 milliseconds to 30 seconds. Its chemical homologue is element number 74, tungsten (W), located just above it in the Periodic Table. Alpha particles (α radiation) are produced from its SF and are used in detecting and measuring its properties. Only one or two atoms are produced at any one time, and they last for only seconds, so determining all of this element's chemical and physical properties is very difficult.

The reaction used by the Berkeley group to create Unh follows:

$$\text{Cf-249} + \text{O-18} + 4n = \text{Unh-106} + \alpha \rightarrow (\text{decays to}) \text{ Unq-259}.$$

HISTORY
Again there is a dispute regarding who discovered element 106 and who should name it. The Russian group at Dubna claimed to have discovered it in June of 1974 by using a cyclotron to shoot heavy Cr-54 + ion particles into atoms of the isotope of Pb-206, to produce element 106. Others could not confirm their results. In the meantime, the Lawrence Livermore laboratories at Berkeley claimed they discovered element 106 in September 1974. The IUPAC, and a group of "referees," decided to name it *unnilhexium* (Unh), which follows the international scheme for naming transactinide elements. In March of 1994 the ACS agreed to name 106 *seaborgium* after Glenn T. Seaborg, who discovered more than ten heavy elements, including plutonium. The IUPAC also proposed the name *seaborgium* (sg), for element 106.

COMMON USES
None, except for research purposes.

EXAMPLES OF COMMON COMPOUNDS
None have been reported, but there may be some related to its homologue, tungsten.

HAZARDS
Same as for any radioactive isotope, but since only a few, short-lived atoms are produced there is no danger to the public.

❖ ❖ ❖

UNNILSEPTIUM (Transactinide Elements)
Interpreted as un = 1, nil = 0, sept = 7, and ium = the Greek suffix for chemicals. The IUPAC's proposed name is bohrium, to honor the Danish physicist/chemist Niels Bohr.
SYMBOL: Uns and Bh   **ATOMIC NUMBER:** 107

ATOMIC WEIGHTS: 261, 262 (?), and 264 **COMMON ISOTOPES:** Three have been produced, 261, 262, and 264.

PROPERTIES AND CHARACTERISTICS
Only a few atoms of element 107 were produced by the following "cold" fusion process:

$$\text{Bi-209} + \text{Cr-54} + \text{neutron} = \text{Uns-262} + \alpha \rightarrow \text{decay chain}$$

Note: Bi-209 is an isotope of bismuth[83], and Cr-54 is an isotope of chronium[24], plus a neutron. The specific chemical and physical properties and characteristics of element 107 have not yet been identified.

HISTORY
In 1976 the nuclear research group in Dubna, Russia, claimed that element 107 could be produced by using an isotope of lead (Pb-208) plus neutrons and an isotope of iron to produce a cold reaction that would produce just a few atoms of 107. They named 107 *nielsbohrium* (Ns). Later, in 1981, a research group (GSI) in Darmstadt, Germany, used a similar reaction in their Separator for Heavy Ion Reaction Products (SHIP) to produce and identify not only element 107, but also 108 and 109. The Darmstadt group, led by P. Armbruster, also decided to name 107 *nielsbohrium*, which was accepted by the IUPAC and the ACS in 1994. Later the Naming Commission of the IUPAC proposed the name *bohrium* for element 107, which is not yet official.

COMMON USES
None, except for research purposes.

EXAMPLES OF COMMON COMPOUNDS
None are known.

HAZARDS
Even though element 107 is an alpha emitter, there is little radiation hazard, since only a few atoms have been produced.

❖   ❖   ❖

UNNILOCTIUM (Transactinide Elements)
May be interpreted as 1-0-8 + ium (as the Greek suffix for chemical). It is also known as hassium.
SYMBOLS: Uno and Hs   **ATOMIC NUMBER:** 108

ATOMIC WEIGHTS: 263, 264, 265, 267, and 269   **COMMON ISOTOPES:**
Five have been identified, 263, 264, 265, 267, and 269.

PROPERTIES AND CHARACTERISTICS
Like element 107, Unniloctium is produced, observed, and identified on the basis of only a few atoms by the following process:

Pb-209 + Fe-58 + neutron = Uno-265 + α → decay products.

Note: Pb-209 is a heavy isotope of lead[82], Fe-58 is an isotope of iron which is bombarded by the lead particles. As with element 107, not much is known about the chemical and physical properties of unniloctium (hassium).

HISTORY
The same group in Darmstadt, Germany, that discovered element 107 discovered 108. Its name, unniloctium, follows the international naming system. The Germans named it *hassium*, which is the Latin name for the province of Hesse in Germany, where their laboratory is located. Both Germany and the ACS have accepted the name *hassium*, and the Naming Commission of the IUPAC also proposed the name.

COMMON USES
None, except for research purposes.

EXAMPLES OF COMMON COMPOUNDS
None have been reported.

HAZARDS
Because only a few atoms are produced at a time, radiation hazard is not a problem.

UNNILENNIUM (Transactinide Elements)

May be interpreted as un = 1, nil = 0, enn = 9 + ium (the Greek suffix for chemical element). It is also know as meitnerium.

SYMBOLS: Une and Mt   **ATOMIC NUMBER:** 109

ATOMIC WEIGHTS: 266 and 268   **COMMON ISOTOPES:** Only two have been reported, 266 and 268.

PROPERTIES AND CHARACTERISTICS

Unnilennium was produced by the following reaction:

Bi-209 + Fe-58 + neutron = Une-266 + $\alpha \rightarrow$ decay chain

Note: Bi-209 is an isotope of bismuth[83], Fe-58 is an isotope of iron[26], Une-266 is the isotope for unnilennium[109], α is for the nuclei of helium atoms, and $\rightarrow$ indicates a decay process to other particles from spontaneous fission.

HISTORY

Element 109, like elements 107 and 108, was synthetically produced, and its existence was confirmed by Peter Armbruster and Gottfried Munzenberg and their research group in the nuclear laboratory located in Darmstadt, Germany. In 1982, they proposed the name *meitnerium*, in honor of Austrian-born physicist Lise Meitner, who was one of the "founders" of nuclear fission.

In 1992 both the American Chemical Society (ACS) and the IUPAC accepted the name *meitnerium* for element 109, while the international name *unnilennium* is still used.

COMMON USES

None, except for research purposes.

EXAMPLES OF COMMON COMPOUNDS

None are known.

HAZARDS

None, except for radiation, which is not much of a risk since only a few atoms are produced at a time.

UNUNNILIUM (Transactinide Elements)

May be interpreted as un = 1, un = 1, nil = 0 + ium (the Greek suffix for chemical elements). Elements ranging from 110 up have not yet received names for individuals or geographic locations, but they do follow the international naming system created by the IUPAC.

SYMBOL: Uun **ATOMIC NUMBER:** 110

ATOMIC WEIGHTS: 267 (?), 269, 271, and 273 (?) **COMMON ISOTOPES:**
There are a number of isotopes of element 110, but not all of them have been
confirmed. The most convincing data is for isotopes Uun-269 and Uun-271.

PROPERTIES AND CHARACTERISTICS

The production and confirmation of elements 110 and elements of higher
atomic numbers required the development of new equipment. Therefore
there was about a ten-year delay in the production, detection, and confirmation of these heavy transactinides. Their chemical and physical properties are not well known. Only a few atoms have been produced, which last
for only a fraction of a millisecond before decaying spontaneously into
particles and radiation.

HISTORY

Germany, Russia, and the United States were all involved in the synthetic
production of element 110 and those beyond it. This search has been an
international effort by a group led by P. Armbruster, who claimed to have
discovered 110 in the Darmstadt, Germany, laboratory in 1994. The isotope
they produced had an atomic weight of 267.

In 1991, A. Ghiorso and others of the Berkeley group produced several
atoms of element 110 and identified them as decay chain products using a new
gas-filled Small Angle Separator System (SASSY-2). The reaction follows:

$$Bi\text{-}209 + Co\text{-}59 + neutron = Uun\text{-}267 + \alpha$$

In 1994 and 1995 S. Hoffman and others used the Separator for Heavy Ion
Reaction Products (SHIP) at the GSI laboratory in Darmstadt to produce
two new isotopes of element 110.

$$Pb\text{-}208 + Ni\text{-}62 + neutron = Uun\text{-}269 + \alpha$$
$$Pb\text{-}208 + Ni\text{-}64 + neutron = Uun\text{-}271 + \alpha$$

Only a few events occurred that produced these isotopes.

A combined team composed of scientists from the Lawrence-Livermore
National Laboratory (LLNL) in Berkeley, California, and from the laboratory in Dubna, Russia, reported the following "hot" fusion reaction:

$$Pu\text{-}244 + S\text{-}34 + neutron = Uun\text{-}273 + \alpha$$

Only one event of this reaction was detected, so it was not confirmed.

COMMON USES

None are known.

EXAMPLES OF COMMON COMPOUNDS
None have been reported.

HAZARDS
None, beyond minor risks from radiation. None of the isotopes of element 110 exist beyond a fraction of a millisecond before they decay into other particles and radiation.

❖   ❖   ❖

UNUNUNIUM (Transactinide Elements)
May be interpreted as un = 1, un =1, un = 1 + ium (the Greek suffix for chemical elements). The IUPAC naming committee has not arrived at a name for unununium.
SYMBOL: Uuu   **Atomic Number:** 111

ATOMIC WEIGHT: 272   **COMMON ISOTOPES:** Only one reported, Uuu-272.

PROPERTIES AND CHARACTERISTICS
Not much is known about element 111, but it is assumed that it is a metal and has some of the characteristics of its homologues, gold and silver, located above it in Group 11 (IB) in the Periodic Table. The reaction that produced unununium follows:

Bi-209 + Ni-64 + neutron = Uuu-272 + $\alpha \rightarrow$ decay

Only three events (atoms) have been detected.

HISTORY
Shortly after element 110 was discovered in 1994, the same group working at the GSI facility in Germany discovered element 111 using the SHIP detection apparatus. One might speculate that the ancient alchemist would be impressed by the production of element 111 by the **transmutation** of two different elements. Very high-speed nickel atoms are made to bombard bismuth atoms. This causes the nuclei of the two atoms to fuse together, producing unununium. Gold, located in Period 6, is unununium's homologue. Unfortunately, only three atoms of element 111 have been produced, and their half-lives were extremely short. These unstable atoms decay into other particles, so transmutated gold is still an alchemist's dream.

COMMON USES
None, except for research.

EXAMPLES OF COMMON COMPOUNDS
None have been reported.

HAZARDS
None, except for minor radiation from just the three atoms ever produced.

❖ ❖ ❖

UNUNBIIUM (Transactinide Elements)
May be interpreted as un = 1, un = 1, bi = 2 + ium (the Greek suffix for chemical elements). No other name has yet been proposed by the IUPAC.
SYMBOL: Uub   **ATOMIC NUMBER:** 112

ATOMIC WEIGHT: 277   **COMMON ISOTOPES:** Only one produced so far, ununbiium-277.

PROPERTIES AND CHARACTERISTICS
The reaction that produced element 112 is as follows:

Pb-208 + Zn-70 + neutron = Uub-277 + $\alpha \rightarrow$ decay particles

Ununbiium is considered a metal because its homologues—mercury, cadmium, and zinc—are located above it in Group 12 (IIB) of the Periodic Table. Not much is know about element 112's properties.

HISTORY
The same international group of scientists that discovered elements 107, 108, 109, and 110—led by P. Armbruster at the GSI in Darmstadt, Germany—discovered element 112. In February of 1996 they announced the synthetic production of one atom of element 112. With the development of sensitive detection equipment, just one atom of Uub was adequate to confirm its fleeting existence. The researchers accelerated zinc atoms to high speeds in a particle accelerator and directed them to smash into atoms of lead. The combined protons in the nuclei of the two atoms (30 in Zn and 82 in Pb) equal the 112 protons in the synthetic atom of ununbiium which, in just a fraction of a millisecond, decays into smaller particles.

COMMON USES
None, except for research.

EXAMPLES OF COMMON COMPOUNDS
None have been reported.

HAZARDS
None.

THE FUTURE TRANSACTINIDE ELEMENTS

Dr. Glenn T. Seaborg of the University of California, who is the discoverer of many heavy elements, recently suggested that since elements from 106 up to and including 112 complete the $6d$ suborbital transition series, with homologues through mercury, it would be difficult to predict what would happen when the next suborbital, $7p$, becomes involved. He suggests that other suborbitals may also become involved in super-heavy elements, and that most likely these new elements will be volatile liquids or gases.

Dr. Darleane C. Hoffman, in her 1996 paper "The Transuranium Elements: From Neptunium and Plutonium to Element 112," presented to the NATO Advanced Study Institute on "Actinides and the Environment," provides the rationale for the possibility of discovering future heavy transactinide elements, including ununtriium (Uut113), ununquadium (Uuq114), ununpentium (Uup115), ununhexium (Uuh116), ununseptium (Uus117), and ununoctium (Uuo118). By using "cold and hot fusion" techniques to produce and identify elements 110 and 111, it was possible to use a similar reaction to produce element 112 (Uub), which was accomplished in 1996. It is expected that these techniques can possibly produce elements 113 and 114 in the same way. The planned reaction follows:

Zn-70 or Ge-76 combining with lead or bismuth, with elements 113 (Uut) and 114 (Uuq) resulting

Another proposed reaction:

Pu-244 + Ca-48 + 4 neutrons = element 114 (Uuq-288) + radiation

Using similar reactions, but improved and more sensitive equipment, it is expected that elements 115, 116, 117, and 118 can be produced in the future. One example of a possible reaction:

Cm-248 + Ca-48 + 4 neutrons = element 116 (Uuh-292) + radiation

This element would have a short half-life.
Note: Cm-248 is an isotope of californium and Ca-48 is an isotope of calcium.

At the time her article was written, Dr. Hoffman said,

"With the expected efficiency of the BGS (Berkeley Gas-filled Separator), a cross section . . . should result in the detection of nearly two events [atoms produced] per week. Thus it now appears possible to reach the long sought superheavy elements."

THE SUPERACTINIDE ELEMENTS

In addition to the above group of artificial transactinide elements, a group of very heavy **superactinide** radioactive elements have been predicted with atomic numbers ranging from 119 to as high as 168 or even 184. As of this writing, they have not been synthetically produced. But if and when they are artificially produced, they will no doubt be radioactive and unstable, and have very short half-lives.

When new elements with atomic numbers beyond 118 are made by man, they will surely be fleeting in existence and only a few atoms in number, but assuredly they will follow the general schema of the Periodic Table of the Chemical Elements.

As Albert Einstein said, "The eternal mystery of the world is its comprehensibility."

GLOSSARY OF TECHNICAL TERMS

abrasive. A finely divided, hard, refractory material used to reduce, smooth, clean, or polish the surfaces of other, less hard substances.

absolute zero. $-273.13°$ Celsius, or $-459.4°$ Fahrenheit, is the coldest temperature possible. It is the temperature at which all matter, such as a gas, possesses no thermal energy, and at which all molecular motion ceases. It has never been reached. At this temperature, "entropy," or the state of disorder, no longer exists. This is an important concept for the third law of thermodynamics and quantum physics. In 1848 Lord Kelvin developed the Kelvin scale for measuring low temperatures, which eliminated the need for negative numbers. Each Kelvin equals one degree Celsius, so zero Kelvin is absolute zero. The freezing point of water is 273 Kelvin ($0°$ Celsius and $32°$ Fahrenheit), and the boiling point of water then becomes 373 Kelvin.

absorption. In chemical terminology, the penetration of one substance into the inner structure of another. Also, the ability of the three forms of matter, i.e., solids, liquids, and gases, to "mix" and intersperse their atoms and/or molecules within the atoms or molecules of other or similar forms of matter.

acid. A substance that releases hydrogen ions when added to water. Strong acids, such as HCl, are proton donors and tend to dissociate in water and become ionized in solution; i.e., they yield H+ ions. They are sour to taste, turn litmus paper red, and react with some metals to release hydrogen gas, e.g., $2HCl + 2Zn = 2ZnCl + H_2$.

actinides. The elements from actinium (atomic number 89) to lawrencium (atomic number 103). They are radioactive, and those beyond uranium (atomic number 92) are produced artificially, although neptunium and plutonium have recently been found in minute quantities in nature. They are all similar to Ac^{89}.

adsorption. Adherence or collection of atoms, ions, or molecules of a gas or liquid to the surface of another substance, called the adsorbent; e.g., hydrogen gas collects (adsorbs) to the surface of several other elements, particularly metals. An important process for dyeing fabrics. Not to be confused with absorption, which is internal mixing or dispersion of one substance within another.

alkali Earth metals. Those metals found in Group 2 (IIA) of the Periodic Table. They are beryllium (Be^4), magnesium (Mg^{12}), calcium (Ca^{20}), strontium (Sr^{38}), barium (Ba^{56}), and radium (Ra^{88}).

alkali metals. Those metals found in Group 1 (IA) of the Periodic Table. They are hydrogen (H^1), lithium (Li^3), sodium (Na^{11}), potassium (K^{19}), rubidium (Rb^{37}), cesium (Cs^{55}), and francium (Fr^{87}).

allotropes. Elements that exist in two or more different forms in the same physical state. The chemical properties of each allotrope are the same, but the physical properties are different; e.g., carbon has four allotropes: diamond, graphite, amorphous carbon, and fullerene. Although the chemical properties of allotropes of an element are the same, some physical properties, such as density, hardness, electrical conductivity, color, and even molecular structure, may differ. The concept of allotropes in chemistry is somewhat similar to the concept of different species of the same plant or animal in biology.

alloy. A mixture of two or more metals, or a mixture of metals and nonmetals. An alloy is a substance made by joining (fusing) a metallic substance with two or more other metals or nonmetals which results in a mixture that has properties different from each of the former substances. Steel is an alloy of iron and carbon. If a small amount of nickel is added, it becomes stainless steel. Brass and bronze are alloys of copper and zinc and copper and tin. Sometimes other metals are also added.

alpha particle. A nucleus of a helium atom (H^{++}), i.e., two positive protons and two neutrons, without any electrons. Such positive particles are produced in nuclear reactions and result from the decay of radioactive elements. They are one of the three basic forms of radiation, i.e., alpha, beta, and gamma. Alpha particles, the heaviest of the three, have a range of penetration in air of five centimeters.

amalgam. A "solution" of various metals that can be combined without melting them together; e.g., gold, silver, platinum, uranium, copper, lead, potassium, and sodium will form amalgams with mercury. Amalgams are alloys that contain mercury.

amorphous. Noncrystalline, having no molecular lattice structure which is characteristic of the solid state; e.g., all liquids and some powders are amorphous. A state of matter having no orderly arrangement of its particles. Glass is an example of a solid amorphous solution.

anesthetic. A chemical compound that induces loss of sensation in a specific part or all of the body.

anhydrous. A substance that contains no water.

anion. An ion having a negative charge that is attracted to the anode (positive pole) in an electrolyte during the process known as electrolysis.

anode. A positively charged electrode, as in an electrolytic cell, storage battery, or electron tube. The anode is where oxidation takes place when electrons are lost in the process.

antiknock agent. A substance added to engine fuel to prevent knocking, i.e., a shuddering sound in engines caused by the uneven and premature burning of fuel.

antimatter. See *antiparticle*.

antioxidant. An organic compound added to rubber, fats, oils, food products, and lubricants to retard oxidation that causes deterioration and rancidity. Also, some foods and diet supplements that neutralize free radicals, such as the hydroxyl free radical (-OH) in human cells.

antiparticle. A subatomic particle, namely a positron, antiproton, or antineutron, with the identical mass of the ordinary particle to which it corresponds but opposite in electrical charge or in magnetic moment. Antiparticles make up antimatter, the mirror image of the particles of matter that make up ordinary matter as we know it on Earth. This is a theoretical concept devised to relate relativistic mechanics to the quantum theory. Some short-lived antielectrons (positrons) were produced in 1932, and antiprotons and antineutrons were seen in particle accelerators in 1955.

aqua regia. A very corrosive, vaporizing yellow liquid that is a mixture of 1 part nitric acid with 3 or 4 parts hydrochloric acid. The ancient alchemists named it aqua regia (meaning "royal water") because it is the only acid that will dissolve platinum and gold.

asphyxiant. A gas that has little or no positive effect but can cause unconsciousness and death by replacing air, thus depriving an organism of oxygen, e.g., carbon dioxide.

atom. The smallest part of an element that can exist which is recognizably part of a particular element. A stray electron could belong to any element, but an atom, with its unique configuration of neutrons (neutral charge), protons (positive charge), and electrons (negative charge), can always be identified as a specific element.

atomic mass (atomic weight). The average (mean) weight of all the isotopes of each specific chemical element found in nature. Atomic *weight* is expressed in atomic mass units (amu), which is defined as exactly 1/12 the *mass* of the carbon-12 atom. It is the total weight of protons plus neutrons found in the nucleus of an atom.

atomic number (proton number). The number of positively charged protons found in the nucleus of an atom, upon which its structure and properties depend. This number determines the location of an element in the Periodic Table. For a neutral atom the number of electrons equals the number of protons.

atomic weight. The total number of protons plus neutrons in an atom. (See *atomic mass*.)

Babbitt metal. A soft alloy of metals such as tin, silver, arsenic, and cadmium combined with a lead base. It can be cast or used as a coating on steel bearings to form an oil-like thin coating which reduces friction. Used to make oil-less bearings.

base. An alkali substance that reacts with (neutralizes) an acid to form a salt, e.g., $4HCl + 2Na_2O \rightarrow 4NaCl + 2H_2O$ [hydrochloric acid + sodium hydroxide yields sodium chloride (table salt) + water]. A base in water solution tastes bitter, feels slippery or soapy, turns litmus paper blue, and registers above 7 on the pH scale. A base can accept two electrons. In water, a base dissociates to yield hydroxyl ions (-OH). See pH.

beta particle. In essence, a beta particle is a high-speed electron and is considered one of the three basic types of radiation—alpha, beta, and gamma. It is a negative particle expelled from the nucleus during the decay of a radioactive atom. If it has a negative charge, it is identical to an electron. If it has a positive charge, it is called a positron (see *antiparticle*). It can travel at up to 99% of the speed of light and can be stopped by several layers of paper.

binary compound. A compound formed by two elements, such as HCl (hydrogen chloride), NaCl (sodium chloride), CO (carbon monoxide), and so forth.

brazing. A welding method in which a nonferrous filler alloy is inserted between the ends or edges of the metals to be joined.

breeder reactor. A nuclear reactor that uses rare U-235, which is fissionable, to change non-fissionable U-238 into Pu-239, which is more abundant and useful for nuclear reactors. A breeder reactor produces about 100 times more fissionable fuel than it starts out with.

carbon dating. Using the rate of decay of carbon-14, a naturally occurring radioactive form of carbon with a half-life of 5,580 years, to date very old organic matter. It can be used to accurately calculate and confirm the date when an organic substance was living by comparing the amount of carbon-14 with carbon-12 remaining in the substance.

carcinogen (carcinogenic). A substance that causes cancer.

catalyst. Any substance that affects the rate of a chemical reaction without itself being consumed or undergoing a chemical change. There are three characteristics of catalysts: (1) Not much is needed to affect the chemical reaction. (2) Although involved in the reaction, a catalyst is not changed or used up. (3) Each catalyst is unique for each specific reaction. Most catalysts accelerate change (positive catalysts), but a few can retard change (negative catalysts). Both types change the amount of energy and/or time required for the chemical reaction to take place. An example is platinum/palladium pellets in automobile catalytic converters that change toxic, harmful gases to less toxic gases, and thus help control pollution. An enzyme is a biological catalyst that affects chemical reactions in living organisms.

catalytic converter. A device used in gasoline engines (internal combustion engines) whereby harmful gases are converted to less harmful gases as they pass through platinum converters. The CO_2 produced is not changed. The platinum/palladium pellets in the converters will last as long as the car since they are not consumed in the chemical reaction. (See also *catalyst*.)

cathode. The negative electrode or plate in an electrolytic system where electrons are gained and reduction takes place. Within the electrolyte, the positively charged ions, called cations, flow to the cathode during electrolysis.

cations. Positively charged ions that migrate to the cathode or negatively charged electrode in an electrolytic system. During electrolysis, cations are attracted to the cathode. Cations are usually positive ions of metals.

cell. *In chemistry*: A cell consists of two electrodes or plates in a liquid electrolyte that provides the pathway for electrons to migrate from one electrode to the other. *In biology*: The smallest unit of a biological structure or an organism capable of independent functioning. It is composed of cytoplasm, one or more nuclei, a number of organelles, and inanimate matter, all of which is surrounded by a semipermeable plasma membrane.

chain reaction (nuclear). Atomic nuclei are bombarded with neutrons and split to release more neutrons, a phenomenon known as fission. These neutrons then split further nuclei, with the release of more neutrons. The process keeps repeating itself, with the generation of enormous amounts of energy. Controlled nuclear chain reactions are the source of energy produced in nuclear power stations, while the nuclear (atomic) bomb must obtain "critical mass" to create an explosion. It takes 33 pounds of uranium-235, or just 10 pounds of plutonium-239, to attain a critical mass that can produce an uncontrolled fission chain reaction resulting in a nuclear explosion.

chemical bond. Electrostatic force that holds together the elements that form molecules of compounds. This attractive force between atoms is strong enough to hold the compound together until a chemical reaction causes the substance to either form new bonds or break the bonds that form the molecule.

chemical reaction. A chemical change involving either endothermic (heat-absorbing) or exothermic (heat-releasing) reactions. Some common examples are oxidation, reduction, ionization, combustion, and hydrolysis.

chemistry. The science concerned with the physical and chemical properties, composition, structures, and interactions of the elements of the Earth. There are many branches of chemistry, including: (1) *analytical chemistry*, the study of "what it is" and "how much there is of it"; (2) *biochemistry*, the study of chemical reactions in living things; (3) *electrochemistry*, the science of the relationships of electricity and chemical changes; (4) *geochemistry*, the study of the composition of the Earth, water, atmosphere, and our natural resources; (5) *industrial chemistry*, or chemical engineering, concerned with the production and use of our resources; (6) *inorganic chemistry*, the study of all atoms and molecules except carbon; (7) *organic chemistry*, the study of carbon compounds and the chemistry of all living things; (8) *physical chemistry*, concerned with the physical properties of matter; and (9) *polymer chemistry*, the study of very large molecules such as rubber and plastics. There are other branches and subbranches of chemistry.

chlorofluorocarbons (CFCs). Compounds containing chlorine, fluorine, and carbon. They are used as refrigerants and as propellants in aerosols. In the upper atmosphere, they decompose to give chlorine atoms, which react with ozone molecules to form chlorine monoxide. Chlorine from the oceans also affects the ozone layer. Such a phenomenon may cause thinning of the ozone layer. They are also suspected, along with CO_2, of increasing the so-called greenhouse effect. Efforts are underway to reduce the use of CFCs.

chlorohydrocarbons (CHCs) and chlorofluorohydrocarbons (CFHCs). Similar to CFCs, except they contain hydrocarbon molecules instead of just carbon. (See also *chlorofluorocarbons*.)

chromatography. Any of a group of techniques used to separate complex mixtures, i.e., vapors, liquids or solutions, by a process of selective adsorption (not to be confused with absorption), the result being that the distinct layers of the mixture can be identified. The most popular techniques are liquid, gas, column, and paper chromatography.

citric acid cycle; TCA cycle; Krebs cycle. An important series of organic enzyme reactions occurring in higher living plants and animals. The cycle begins and ends with oxaloacetic acid. This cycle is necessary for the oxidation and metabolism of glucose (sugars) to release energy. This cycle is the major pathway for oxidation which takes place in animals, plants, and some bacteria. The TCA cycle is thought to predate all life on Earth and provided the necessary chemical reactions that paved the way to form amino acids and, ultimately, life. In England in 1953, Hans A. Krebs (1900–1981) was awarded the Nobel Prize for physiology and medicine for describing the citric acid cycle which bears his name.

cladding. The process by which two different metals are rolled together under heat and pressure to form a bond between them. The interface between the two metals becomes an alloy, but the "outerface" of each maintains its original properties. An example is when low-grade steel is "clad" with high-grade stainless steel or some other noncorrosive metal.

combustion. A chemical reaction in which a substance combines with a gas to give off heat and light (fire). It commonly involves an oxidation reaction which is rapid enough to produce heat and light.

compound. A substance in which two or more elements are joined by a chemical bond to form a substance different from the combining elements. The combining atoms do not vary their ratio in the new compound and can only be separated by a chemical reaction, not a physical force.

condensation. Change of a gas or vapor into a liquid by cooling.

conductors. Substances that allow heat or electricity to flow through them.

control rods. Rods made of metals that have a high capacity to absorb neutrons in a nuclear reactor. Their purpose is to control the rate of fission in nuclear reactors since they absorb the neutrons that sustain chain reactions.

convulsion. A violent spasm.

corrosion. The electrochemical degradation of metals or alloys due to reaction with their environment, which is accelerated by the presence of acids or bases. Also, the destruction of body tissues by strong acids and bases. The rusting of iron is a form of corrosion called oxidation.

covalent bond (homopolar). Sharing of electrons by two or more atoms. A single covalent bond involves the sharing of two electrons with two different atoms. In a double covalent bond, four electrons are shared, and in a triple bond, six electrons are shared. In a nonpolar covalent bond, the electrons are not shared evenly, while in a polar bond, they are definitely shared evenly.

cracking. The process used in the refining of petroleum, by which hydrocarbon molecules are broken down into smaller molecules, which are then used in the production of chemicals and fuels. Heat was originally used to crack petroleum, but today most hydrocarbon fuels are formed by catalytic cracking. Platinum pellets or silica alumina is usually used as the catalyst.

critical mass. The minimum mass of fissionable material (U^{235} or Pu^{239}) that will initiate an uncontrolled fission chain reaction, as in a nuclear (atomic) bomb. The critical mass of uranium-235 is about 33 pounds, and of plutonium-239, about 10 pounds.

crucible. A cone-shaped container having a curved base and made of a refractory (heat-resistant) material, used in laboratories to heat and burn substances. Crucibles may or may not have covers. A crucible is also a special type of furnace used in the steel industry which has a cavity for collecting the molten metal.

cryogenic. Study of the behavior of matter at very low temperatures below $-200°C$. The use of the liquefied gases—oxygen, nitrogen, and hydrogen—at approximately $-260°C$ is standard industrial practice. (See also *superconductor*.)

crystalline. A substance that has its atoms, molecules, or ions arranged in a regular three-dimensional structure, e.g., igneous or metamorphic rock and diamonds.

cupric. Form of the word copper used in naming copper compounds in which the copper has a valence of +2, e.g., cupric oxide (CuO) and cupric chloride ($CuCl_2$).

cuprous. Form of the word copper used in naming copper compounds in which the copper has a valence of +1, e.g., cuprous oxide (Cu_2O), and cuprous chloride (CuCl).

cyclotron. A machine designed to accelerate the movement of subnuclear particles to explore the nature of matter by "smashing" them into atoms.

decomposition. The process whereby compounds break up into simpler substances. Decomposition usually involves the release of energy as heat.

dehydration. The removal of water, usually 95% or more, from a substance or a compound by exposure to a high temperature. Also, the excessive loss of water from the human body.

dehydrogenation. A chemical process that removes hydrogen from a compound. It is considered a form of oxidation.

deliquescence. The absorption of water from the atmosphere by a substance to such an extent that a solution is formed, i.e., water-soluble chemical salts (powders) which dissolve in the water absorbed from the air.

density. The relative measure of how much material is in a substance as compared to a standard. It may be calculated by dividing weight (mass) by volume. Density of solids and liquids is compared to the density of 1 milliliter (ml or cc) of water at 4°C. The weight of 1 cc of water is exactly 1 gram, thus the density of water = 1. Therefore, water is used as a relative standard for comparison of the densities of solids and liquids (weight/volume or g/cc). For gases, the density is expressed as grams per liter (g/L); for air, it is 1.293. Thus, the specific gravity of air = 1.0 (1.293/1.293), and 1.0 is used as the standard for comparing the specific gravity (density) of gases. Specific gravity is the same as density for liquids and solids, but it is not exactly the same as density for gases. (See also *specific gravity*.)

depilatory. A substance (usually a sulfide) used to remove hair from skin of humans or the hides of cattle, pigs, and horses.

depression. A general feeling of uneasiness or distress about past or present conditions or future uncertainties. It can be as mild as just a day or two of "feeling low" to clinical depression, which is a serious disease needing medical and/or psychological treatment.

desiccant. A hygroscopic substance (e.g., silica gel) that can adsorb water vapor from the air. Used to maintain a dry atmosphere in food containers, chemical reagents, and so forth.

deuteron. A nuclear particle, with a mass of 2 and a positive charge of 1. The deuterium atom is called heavy hydrogen (^2D) (hydrogen with an atomic weight of 2).

diatomic molecule. A molecule that contains two atoms of the same element, e.g., H_2, O_2, and N_2.

die. A device with a cone-shaped hole through which soft metal is forced to form wire, pipe, or rods. Also, a form for cutting out or forming materials. (See also *ductile* and *extrusion*.)

diode. An electronic device that restricts the flow of electrical current to one direction, such as a vacuum tube made up of two electrodes, one cathode and one anode, or a semiconductor used as a rectifier to change AC current to DC current.

distillation. A process used to purify and separate liquids. A liquid is said to have been distilled when it has been heated to a vapor and then condensed back into a liquid.

divalent. Having a valence of 2. Similar to bivalent.

ductile. Easily shaped. A ductile substance can be extruded through a die and easily drawn into wires. (See also *die*.)

eka. Eka means "first" in Sanskrit. It was used by Mendeleyev as a prefix for yet-to-be-discovered elements that he predicted would "fit" into specific blank spaces in the next lowest position of the same Groups in his Periodic Table. For example, he knew that the unknown element for the blank space following calcium would be closely related to boron, so he gave it the name of "eka-boron" until the unknown element, scandium, was definitely identified and named. Scandium's properties are almost exactly as Mendeleyev predicted they would be. Several other eka elements were eka-aluminum (for gallium), eka-silicon (for germanium), and eka-manganese, although Mendeleyev missed this one.

electrochemistry. Use of electricity as the energy source to break up the oxides of elements. Used to extract metals from ores.

electrode. A conductor that allows an electric current to pass through an electrolyte (solution). The current enters and leaves the electrolyte solution through anode and cathode electrodes. Electrodes are made of metals such as copper, platinum, silver, zinc, lead, and sometimes carbon.

electroluminescence. Luminescence generated in crystals by electric fields or currents in the absence of bombardment or other means of excitation. It is observed in many crystalline substances, e.g., silicon carbide, germanium, and diamond.

electrolysis. A process in which an electric current is passed through a liquid, known as an electrolyte, producing chemical changes at each electrode. The electrolyte decomposes, thus enabling elements to be extracted from their compounds. Two examples are the production of chlorine gas by the electrolysis of sodium chloride (NaCl, or table salt), and the electrolysis of water to produce oxygen and hydrogen.

electrolyte. A compound that, when molten or in solution, will conduct an electric current. The electric current decomposes the electrolyte. Some salts and acids make excellent electrolytic solutions.

electromagnetic spectrum. Electromagnetic waves, i.e., the total range of the wavelengths of radiation, from the very short high-frequency waves, including cosmic rays, gamma rays, X-rays, and ultraviolet radiation, to wavelengths of visible light (violets to reds), and then the long infrared heat radiation, microwaves, and finally the longest electromagnetic radiation of the spectrum, such as radio waves and electric currents.

electron. An extremely small, negatively charged particle that moves around the nucleus of an atom. An electron's mass is only 1/1837 of the mass of a proton. The number of electrons in a neutral atom is always equal to the number of positively charged protons in the nucleus. Ions are similar to atoms with an electrical charge. For ions, the number of electrons does not equal the number of protons. The electrons surrounding the nuclei of atoms are responsible for the chemical properties of elements. The interaction of the electrons of atoms *is* the chemistry of our Earth's elements.

electronegative. The positive nucleus can exert attraction for additional electrons, even though, as a whole, the atom is electrically neutral. The nuclei of metals have a very weak attraction for additional electrons. Thus, they exhibit weak electronegativity (by giving up electrons to become positive ions), while the nuclei of nonmetals have a tendency to attract extra electrons and, thus, are highly electronegative (gain electrons to become negative ions). When atoms "share" electrons, the atom that "keeps" the electrons for most of the time is the most electronegative partner. The most highly electronegative elements are fluorine, chlorine, oxygen, sulfur, and nitrogen.

electroplating. Sending an electrical charge from one electrode through an electrolyte to deposit a thin coating of a metal on an object at the opposite electrode.

electropositive. The opposite of electronegative, e.g., metals.

electrostatic. Relating to stationary electric charges.

element. A pure substance composed of atoms. Each specific element's atoms have the same number of protons in their nuclei. Elements cannot be broken down into simpler substances by normal chemical means. Each element has a unique proton number and contains an equal number of electrons and protons. Thus, the atoms of elements have no electrical charge. All substances and materials are composed of about one hundred elements. Many of the heavier elements are man-made, unstable, radioactive, and decay by fission into other elements. (See also *atom*.)

emetic. A chemical agent that induces vomiting.

energy. The capacity for doing work. There are many forms of energy. Two classic classifications are potential energy and kinetic energy. The word potential is derived from the Latin word *potens*, which means "having power." The word kinetic is derived from the Greek word *kinetos*, which means "moving."

Several forms of energy are: (1) *mechanical energy*, e.g., using a sledgehammer; (2) *chemical energy*, e.g., burning fuel, rusting or oxidation, and photosynthesis; (3) *electrical energy*, e.g., a generator sending electrons through wires and lightning; (4) *radiant energy*, e.g., sunlight and X-rays; (5) *nuclear energy*, e.g., the fission or fusion of atomic nuclei and $E = mc^2$. There are many sources of energy. All of them except nuclear energy involve the electrons of atoms, viz., chemistry.

enzyme. Any of a number of proteins or conjugated proteins which are produced by living organisms and which act as biochemical catalysts in those organisms.

evaporation. The process whereby a liquid changes to a gas. The "interface" between the liquid state of a substance and the gaseous state of that substance. (See also *liquid* and *gas*.)

extrusion. The process whereby a material is forced through a cone-shaped hole in a metal die, followed by cooling and hardening.

fermentation. A chemical change induced by bacteria in yeast, molds, or fungi which involves the decomposition of sugars and starches to ethanol and carbon dioxide. Used in the preparation of breads and the manufacture of beer, wine, and other alcoholic beverages.

ferromagnetic. Characteristic of iron and other metals and alloys of iron, nickel, cobalt, and various materials, which possess a high magnetic permeability. Also, the ability to acquire high magnetization in a weak magnetic field.

filament. A continuous, finely extruded wire, usually tungsten, gold, or a metal carbide, or synthetic material that is heated electrically to incandescence in electric lamps. Also used to form cathodes in electron tubes.

fission (nuclear). The splitting of an atom's nucleus with the resultant release of enormous amounts of energy and the production of smaller atoms of different elements. Fission ocurs spontaneously in the nuclei of unstable radioactive elements, such as uranium-235 or plutonium-239. This process is used in the generation of nuclear power, as well as in nuclear bombs.

flammable. Easily bursting into flame. Combustible.

fluorescent. Consisting of a gas-filled tube with an electrode at each end. Passing an electric current through the gas produces ultraviolet radiation which is converted into visible light by a phosphor coating on the inside of the tube. This emission of light by the phosphor coating is called fluorescence. (See also *phosphor.*)

flux. A chemical that assists in the fusing of metals while preventing oxidation of the metals, such as in welding and soldering. Also, lime added to molten metal in the furnace to absorb impurities and form a slag that can be removed.

forge. A furnace or hearth where metals are heated and softened (wrought), e.g., wrought iron.

fossil fuel. Fuel resulting from an organic source in past ages that is buried within the earth, e.g., petroleum, natural gas, and coal, and which is considered nonrenewable.

fractional crystallization. The separation of ores by using their unique rates of crystallization. Different minerals crystallize at different temperatures and rates. Thus, they can be separated if these factors are controlled.

fractional distillation. A process used to obtain a substance in a form as pure as possible. The evaporated vapors (fractions) are "boiled" off and collected at different temperatures, which separates the fractions by either their different boiling points or their condensation temperatures.

fractionation. The process of separating and isolating a mixture of either gases or liquids. The separation may be accomplished by distillation, crystallization, electrophoresis, gel filtration or chromatography, particle filtration, or centrifugation. This procedure is based on the fact that the different components of the mixture have different physical and chemical properties.

free electron. An electron not attached to any one atom and not restricted from flowing.

free element. An element found in nature, uncombined with other elements.

free radical. A molecular fragment having one or more unpaired electrons and charges, usually short-lived and highly reactive, e.g., the hydroxyl radical, -OH, which is thought to contribute to the aging process of our bodies.

fuel cell. A cell in which the chemical energy from oxidation of a gas (fuel) is directly converted into electricity. An electrochemical device for continuously converting chemicals (a fuel, such as H_2, and an oxidant, such as O_2) into direct-current electricity and water. Once the device is operating, a "redox" or reduction + oxidation reaction takes place releasing electrical energy. Also, an aircraft fuel tank made of or lined with an oil-resistant synthetic rubber.

fungicide. A chemical compound or solution that will destroy or inhibit the growth of fungi.

fusion (thermonuclear reaction). An endothermic nuclear reaction yielding large amounts of energy in which the nuclei of light atoms, such as deuterium and tritium, join or fuse to form helium, e.g., the energy of our sun and the hydrogen bomb. (The opposite of fission.)

galvanization. The process of coating iron or steel with a layer of zinc, making it more resistant to corrosion than the underlying metal.

gamma rays. A type of radiation emitted by radioactive elements. Gamma rays are capable of traveling through 15 centimeters of lead. Because of their ability to kill body cells, they are used in the treatment of cancer. Gamma rays have the shortest wavelength and most energy of the three major types of radiation, i.e., alpha, beta, and gamma.

gas. A state of matter in which atoms and molecules are linked by very few bonds and are, therefore, given great freedom of movement. Gases have a fixed mass (weight), but no fixed shape or volume.

gastrointestinal. Relating to the stomach and intestines.

Geiger counter. Also known as a scintillation counter. A device used to detect, measure, and record radiation. The instrument gets its name from one of its parts, the Geiger tube, which is a gas-filled tube containing coaxial cylindrical electrodes.

getter. Also called a scavenger. Any substance added to a mixture of substances used to reduce or inactivate any traces of impurities. A metal added to alloys to reduce impurities in the alloys. Also, small amounts of metals inserted into vacuum tubes to absorb any remaining trace gases from the vacuum. (See also *misch metal*.)

graphite. The crystalline allotrope of carbon, which is steel gray to black. Used in lead pencils, lubricants, and coatings. Also fabricated forms, such as molds, bricks, crucibles, and electrodes. Used to moderate radiation in nuclear reactors.

Haber process. A special process for making ammonia from hydrogen and nitrogen. Named after Fritz Haber (1868–1934) who, along with Carl Bosch (1874–1940), developed the very important ammonia-making process.

half-life. The time required for one-half of the atoms of a radioactive element to decay or disintegrate by fission.

halides. Binary compounds of the halogens, i.e., fluorides, chlorides, iodides, and bromides, e.g., LiF, NaCl, KI, and so forth.

halogens. Electronegative monovalent nonmetallic elements of Group 17 (VIIA) of the Periodic Table (fluorine, chlorine, iodine, bromine, astatine). In pure form, they exist as diatomic molecules, e.g., Cl_2. Fluorine is the least dense and most active, while iodine is the most dense and least active (excluding radioactive astatine).

heat. The energy produced by the motion of molecules. Heat may be thought of as the energy that is transferred from one body to another by one of the following three forms of heat transfer: radiation, conduction, or convection. Since all molecular motion ceases at absolute zero, no heat will be evident at $-273.13°C$, or $459.4°F$. Temperature may be thought of as an average of the total molecular motion (heat) one body has that may be transferred to other bodies. Temperature is measured by thermometers and thermocouples which react to heat.

herbicide. A pesticide, either organic or inorganic, used to destroy unwanted vegetation, e.g., weeds, grasses, and woody plants.

homologue. Homogenous (Latin, "the same kind") means the elements in a Group have very similar characteristics and properties. Thus, they react in much the same way.

hydration. Combining with water to form a hydrate, i.e., a compound containing water combined in a definite ratio, the water being retained or thought of as being retained in its molecular state.

hydride. An inorganic compound (either binary or complex) of hydrogen with another element. A hydride may be either ionic or covalent. Most common are hydrides of sodium, lithium, aluminum, boron, e.g., LiH, KH, and AlCHO (a complex hydride compound of aluminum + the carbonal –CHO group).

hydrocarbon. An organic compound whose molecules consist of *only* carbon and hydrogen. Hydrocarbons can form chains of molecules known as "aromatic compounds" or form stable rings known as "aliphatic compounds." All petroleum products and their derivatives are hydrocarbons.

hydrocracking. The cracking of petroleum and/or petroleum products using hydrogen. (See also *cracking* and *hydrocarbon*.)

hydrogasification. Production of gaseous or liquid fuels by direct addition of hydrogen to coal.

hydrogenation. Also known as hardening. It involves the conversion of an unsaturated compound into a saturated compound by the addition of hydrogen, e.g., converting liquid oils to solids by adding hydrogen, such as corn oil to margarine.

hydroxide. A chemical compound containing the hydroxyl group of –OH, such as bases, certain acids, phenols, and alcohols, which form OH$^-$ ions in solution. Most metal hydroxides are bases, e.g., the alkali metals.

hygroscopic. Descriptive of a substance that can absorb water from the atmosphere, but not enough to form a solution.

igneous rock. Rock formed by the cooling of magma (molten rock).

incandescent. Giving off a white light after being heated to a particularly high temperature. For example, household electric light bulbs produce incandescent light due to the electricity heating the bulb's filaments in a vacuum or inert atmosphere.

inert. Unreactive and does not readily form compounds.

inert gases (noble gases). Extremely unreactive gases found in Group 18 (VIIIA) of the Periodic Table.

infrared light. Radiation of wavelengths longer than the visible light wavelength. Invisible to the naked eye. Similar to "heat" rays.

ingested. Eaten and swallowed into the gastrointestinal system.

ingot. A mass of metal that is shaped into a bar or block. Also, a casting mold for metal. Ingot iron is a highly refined steel with a high degree of ductility and resistance to corrosion.

inhaled. Breathed into the lungs.

inorganic. Involving neither organic life nor the products of organic life or compounds of carbon.

inorganic chemistry. The study of all elements other than carbon compounds.

insecticide. A chemical used to kill insects.

insoluble. Cannot be dissolved in a liquid or made into a solution.

insulator. Any substance or mixture that has an extremely low level of thermal or electrical conductivity, e.g., glass, wood, polyethylene, polystyrene.

intoxication. A state that occurs from inhaling industrial fumes, e.g., zinc smelting, or ingesting too much of a substance such as zinc, alcohol, or drugs, the results of which are lethargy, depression, possible addiction, and other health problems.

ion. An atom or group of atoms that have gained or lost electron(s) and, thus, have acquired an electrical charge. The loss of electrons gives positively charged ions. The gain of electrons gives negatively charged ions. If the ion has a net positive charge in a solution, it is a cation. If it has a net negative charge in solution, it is an anion. An ion often has different chemical properties than the atoms from which it was formed.

ionic bond (electrovalent bond). The bond formed by transfer of one or more electrons from one atom to another during a chemical reaction. It is similar to a chemical bond where there is an attraction between atoms that is strong enough to form a new chemical unit (compound) that differs in its properties from the bonding atoms.

isotopes. Atoms of the same element with different numbers of neutrons. All atoms of an element always contain the same number of protons. Thus, their proton (atomic) number remains the same. However, an atom's nucleon number, which denotes the total number of protons and neutrons, can be different. These atoms of the same element with different atomic weights (mass) are called isotopes. Isotopes of a given element all have the same chemical characteristics (electrons), but they may have slightly different physical properties.

lanthanide elements. Also known as rare-Earth elements. Elements in Period 6, starting at Group 3, of the Periodic Table. The lanthanide elements start with proton number 57 (lanthanum, La^{57}) and continue through number 71 (lutetium, Lu^{71}).

laser. A device that produces a continuous single-colored (frequency or wavelength) beam of light. Laser light is very intense, can travel long distances, moves in one direction, and does not spread. Lasers are used in surgery, in supermarket bar code scanners, and in the production of holograms. The word "laser" is derived from the phrase **L**ight **A**mplification by the **S**timulated **E**mission of **R**adiation.

legume. A plant that takes N_2 from the atmosphere and incorporates it into nodules on its roots, e.g., bean and pea plants.

liquid. A state of matter in which the bonds between particles are weaker than those found in solids, but stronger than those found in gases. Liquids always take the shape of the vessels in which they are contained. They have fixed masses and volumes but no fixed shape.

lubricant. An oily substance, in either a liquid or solid form, that reduces friction, heat, and wear when applied as a surface coating to moving parts or between solid surfaces.

luminescence. Emitting light without giving off heat. Also known as "cold light." (See also *phosphor.*)

machinable. The ability to be rolled, pounded, and cut on a lathe.

magnet. A body or an object that has the ability to attract certain substances, e.g., iron. This is due to a force field caused by the movement of electrons and the alignment of the magnet's atoms.

magnetic field. The space around a magnet or an electrical current where the existence of magnetic force is detectable at every point.

malleable. The ability to be beaten or hammered into different shapes. Malleable materials are usually ductile as well.

mass. The quantity (amount) of matter contained in a substance. Mass is constant regardless of its location in the universe. The weight of a body is dependent on the size (mass) of the planet, plus the square of the distance between the object and the planet. The weight of an object on Earth can be thought of as the gravitational attraction between the object and the Earth. The greater the mass of an object, the more it will weigh on the Earth, but its mass will be the same on the moon even though the moon's gravity is 1/6th that of Earth's.

mass number. The total number of protons and neutrons in the nucleus of an atom. Also known as the nucleon number.

matter. Everything that has mass and takes up space, regardless of where it is located in space. Modern physics has shown that matter can be transformed into energy, and vice versa; $E = mc^2$. (See also *mass.*)

melting point. Temperature at which a solid changes to a liquid.

metabolism. A chemical transformation that occurs in organisms when nutrients are ingested, utilized, and finally eliminated, e.g., digestion and absorption, followed by a complicated series of degradations, syntheses, hydrolyses, and oxidations that utilize enzymes, bile acids, and hydrochloric acid. Energy is an important by-product of the metabolizing of food.

metal. An element that is shiny, ductile, malleable, a good conductor of heat and electricity, has high melting and boiling points, and tends to form positive ions during chemical reactions. Metals exhibit very low electronegativity and become cations which are attracted to the negative electrode or cathode in electrolytic solutions. Different metals possess these properties in varying degrees. Most of the 111 elements so far discovered are classified as metals. Most metals, other than those in Groups 1 and 2, are called transition metals. Metals exhibit oxidation states of 1+ (the alkali metals) to 7+ (some of the transition metals.)

metalloid. An element that is neither a metal nor a nonmetal. It has properties that are characteristic of both groups, e.g., silicon, selenium, and arsenic. The term "metalloid," although descriptive, is an old word used for these types of elements, which are now referred to as semiconductors.

metallurgy. The process of extracting and purifying metals from their ores, resulting in the creation of usable items and products.

mineral. An inorganic substance that occurs in nature, e.g., limestone, sand, clay, coal, and many ores.

misch metal. A commercial mixture of rare-Earth metals that are flammable. The mixtures are compacted into various forms. Misch metals create sparks with only slight friction and are used for flints in cigarette lighters and other "spark" igniters. They are also used as "getters" to absorb gases in vacuum tubes and to remove impurities from metal alloys.

moderator. A substance of low atomic weight, e.g., beryllium, carbon (graphite), deuterium, or ordinary water, which is used in nuclear reactors to control the neutrons that sustain chain reactions.

molecule. The smallest particle of an element containing more than one atom (e.g., O_2) or a compound that can exist independently. It is usually made up of a group of atoms joined by covalent bonds.

monatomic molecule. A molecule containing only one atom. Only the noble gases have monatomic molecules.

monovalent (univalent). An atom having a valence of 1.

mordant. A fixative used in the textile industry. A substance that binds or fixes a dye to a fiber so it won't run.

mutation. The process of being changed or altered, i.e., the alteration of inherited genes or chromosomes of an organism.

natural gas. A mixture of hydrocarbon and combustible gases produced within the Earth and extracted through oil and gas wells. Although its composition varies, depending on where it was formed in the Earth, natural gas is composed of about

90% methane (CH_4). It is moved from place to place through extensive pipelines in the United States and can be liquefied for easy storage and shipping.

neutron. A fundamental particle of matter with a mass of 1.009 and having no electrical charge. It is a part of the nucleus of all elements except hydrogen.

neutron absorber. A substance that absorbs neutrons. Used to make control rods in nuclear reactors to adjust the rate of fission by absorbing neutrons.

noble gases. Also called inert gases, found in Group 18 (VIIIA) of the Periodic Table. They are extremely stable and unreactive.

noble metals. Unreactive metals, e.g., platinum and gold, which are not easily dissolved by acids and not oxidized by heating in air. Not a precise descriptive chemical term.

noncombustible. A material, which can be a solid, liquid, or gas, that will not ignite or burn despite extreme high temperatures, e.g., water, carbon dioxide.

nonmetal. A classification of twenty-five elements. One group has moderate electrical and thermal conductivity, e.g., semiconductors (also called metalloids); the other group has very low thermal and electrical conductivity but high electronegativity, e.g., insulators and any material not classed as a metal, such as plastics, oil, paper, and so forth. Nonmetals form ionic and covalent compounds by either gaining or sharing electrons with metals, and thus becoming negative ions. In electrolytic solutions, nonmetals form anions, which are attracted to the positive electrode (anode).

nuclear reactor. An apparatus in which controlled nuclear fission chain reactions take place. The heat produced by such reactions is used to generate electricity. Fissionable elements used in nuclear reactors can be used to make nuclear weapons. Nuclear reactors can also be used to produce numerous radioisotopes.

nucleon number. The total number of protons and neutrons in the nucleus of an atom, which equals the atomic weight (mass) of an atom. (See also *mass number*.)

nucleus. The core of an atom, which provides almost all of the atom's mass. It contains protons and neutrons (except for hydrogen, which has no neutrons) and has a positive charge equal to the number of protons. This charge is balanced by the negative charges of the orbital electrons.

orbital electrons. Electrons found in shells outside the nucleus of an atom.

orbital theory. A quantum theory of matter dealing with the nature and behavior of electrons in either a single atom or combined atoms. The theory describes the part of the atom where electrons are most likely to be found as energy levels.

ore. A metal compound found in nature. Also, a mineral from which it is possible to extract the metal.

organic chemistry. The study of carbon compounds, e.g., alcohols, ethers, foods, animals, and so forth. It does not include the carbonates, hydrogen carbonates, or oxides of carbon. Organic chemistry includes the chemistry of living things and their products as well as carbon chemistry.

oxidation. A reaction in which oxygen combines chemically with another substance and undergoes one of three processes, i.e., the gaining of oxygen, the loss of hydrogen, or the loss of electrons. In oxidation reactions the element being oxidized loses electrons to form positive ions, while in complementary chemical reactions called reduction, electrons are gained to form negative ions.

oxide. A compound formed when oxygen combines with one other element, either a metal or nonmetal, e.g., magnesium oxide, copper oxide, and carbon dioxide.

oxidizing agent. An agent that induces oxidation in substances. The agent itself is reduced or gains electrons, e.g., peroxides, chlorates, nitrates, perchlorates, and permanganates.

ozonosphere (ozone layer). The layer is found in the upper atmosphere, between 10 and 30 miles in altitude. This layer contains a high concentration of ozone gas (O_3), which partially absorbs solar ultraviolet (UV) radiation and prevents it from reaching the Earth. There is some controversy about both natural and man-made chemicals causing a thinning of this protective layer of ozone. The destruction of ozone (O_3 to O_2 and O) by UV and the natural reformation of new ozone from atmospheric oxygen have been in somewhat of a balance for aeons. The question is, can nature continue this balance with the addition of chemicals, both natural and man-made, that affect the ozone system?

paramagnetic. Characteristic of an element in which an induced magnetic field is in the same direction and greater in strength than the actual magnetizing field, but much weaker than in ferromagnetic materials.

particle. A very small piece of a substance that maintains the characteristics of that substance. Subatomic particles are protons, neutrons, and electrons, which are smaller than atoms. Many subnuclear particles have also been discovered.

particle accelerator. A machine designed to speed up the movement of atomic particles and electrically charged subatomic particles that are directed at a target. These subatomic particles are also called "elementary particles" and cannot be further divided. They are used in high-energy physics to study the basic nature of matter and to try to determine the ultimate origin of nature, life, and the universe. Particle accelerators are also used to synthesize elements by "smashing" subatomic particles into nuclei to create new, heavy, unstable elements, such as the superactinides.

petrochemicals. Organic chemical compounds containing carbon and hydrogen which are derived from petroleum or natural gas.

pH. A scale of numbers (1 to 14) used to measure the acidity or alkalinity of a substance or solution. A substance with a *p*H of 1 to 6 is acidic, one with a *p*H of 7 is neutral (e.g., pure water), and one with a *p*H of 8 to 14 is alkaline or basic. Technically it is defined as the logarithm of the reciprocal of the hydrogen+ ion concentration in a solution.

pharmaceutical. Relating to manufacture and sale of drugs as well as medicinal, dietary, and personal hygiene products.

phlogiston. The hypothetical substance believed to be the volatile component of combustible material. It was used to explain the principle of fire before oxidation and reduction were known and prior to the discovery of the principle of combustion by Lavoisier.

phosphor. A substance, either inorganic or organic, or in liquid or crystal form, that absorbs energy and emits light (luminescence). If the substance glows for only a very short time after being excited, it is called "fluorescent," e.g., a fluorescent light tube. If it continues to glow for a long time, it is called "phosphorescent," e.g., the afterglow of a TV screen when it is turned off, and glow-in-the-dark clock faces. (See also *luminescence* and *phosphorescence*.)

phosphorescence. The emission of light after a substance has ceased to absorb radiation; i.e., it glows in the dark. Electrons of the atoms that absorb energy are moved to a higher energy level (shell). After the source of energy is removed, the energized electrons jump back to their original level, or lower shell. In doing so, the excited electrons continue to emit light energy many hours or days later. (See also *phosphor*.)

photoconductors. Substances whose ability to conduct electricity is affected by light. They are poor conductors of electricity in darkness but good conductors in light, e.g., selenium.

photoelectric cells. There are two types of these cells: photoconductive and photovoltaic. Electrical conductivity of photoconductive cells increases as light grows brighter. Used in photographic light meters to control light conditions and correct exposures. Also used in burglar alarms. Photovoltaic cells convert light into electricity.

photon. The quantum unit of electromagnetic radiation or light. According to Albert Einstein's theory, light can be thought of as either waves or particles or both. This concept is difficult to understand because it is described only in mathematical terms. Photons are emitted when electrons are excited and move from one energy level (shell) to another.

photosynthesis. Process by which chlorophyll-containing cells in plants and bacteria convert carbon dioxide and water into carbohydrates, resulting in the simultaneous release of energy and oxygen. The process of photosynthesis is responsible for all food on Earth.

piezoelectricity. The generation of electric energy by the application of mechanical pressure, usually on a crystalline substance such as quartz. The term is named after the Greek word *piezo*, which means "to squeeze."

pigment. A substance, usually a dry powder, used to color another substance or mixture.

pitchblende. A brownish-black mineral of uraninite and uranium oxide with small amounts of water and uranium decay products. The essential ore of uranium.

plasma. An electrically neutral, highly ionized gas, which is made up of ions, electrons, and neutral particles. A plasma is produced by heating a gas to a high temperature, which causes all the electrons to separate from their nuclei, thus forming ions. These ion nuclei can be forced together under great pressure to form larger nuclei and produce tremendous amounts of energy, e.g., nuclear fusion, the sun, the stars, and the hydrogen bomb.

platinum metals. Elements that tend to be found together in nature, e.g., ruthenium, osmium, rhodium, iridium, palladium, and platinum.

polarize. To effect the complete or partial polar separation of positive and negative electric charges in a nuclear, atomic, molecular, or chemical system. Or, to reduce the glare of sunlight by changing the orientation of the light waves passing through the glass (used in windows and eyeglasses).

polymer. A natural or synthetic compound of high molecular weight. Examples of natural polymers: polysaccharide, cellulose, and vegetable gums. Examples of synthetic polymers: polyvinyl chloride, polystyrene, and polyesters.

polyunsaturated. Relating to long-chain carbon compounds, those having two or more double bonds (unsaturated) per molecule, e.g., linoleic and linolenic acids, commonly known as corn oil and safflower oil, respectively.

positron. The positively charged antiparticle of an electron. (See also *antiparticle*.)

prism. A homogeneous, transparent solid, usually with a triangular base and rectangular sides, used to produce or analyze a continuous spectrum of light.

proliferation. The production or reproduction of new growth in a rapid and repeated manner.

proton. A positively charged particle found in the nucleus of an atom. For neutral atoms the number of protons is always equal to the number of negatively charged electrons, and this number is the same as the proton number.

proton number. Also known as the atomic number. The number of protons found in the nucleus of an atom. Every element has a unique proton number, and every atom of that same element has the same proton number. Elements are arranged in order of their proton number in the Periodic Table.

pyrophoric. Any material that can ignite spontaneously in air at just about room temperature, e.g., phosphorus, lithium hydride, and titanium dichloride. Or, a metal or alloy that will spark when slight friction is applied to it, e.g., barium or misch metal (used as flint in cigarette lighters).

pyrotechnics. The manufacture and/or release of fireworks, flares, and warning equipment. Fireworks are composed of oxidizers, such as potassium nitrate and ammonium perchlorate, and fuels, such as aluminum, dextrin, sulfur, magnesium, and so forth. Colors produced by some of the compounds of a few elements that are added to fireworks are: strontium = red; barium = green; copper = blue; sodium = yellow; magnesium and aluminum = bright white.

quantum. The basic unit of electromagnetic energy, equal for radiation of frequency v to the product hv, where h is Planck's constant. The quantum number is the basic unit used to measure electromagnetic energy. To simplify, a very small bit of something. The photon is a quantum amount of electromagnetic radiation. The quantum theory of energy ushered in the era now known as "modern physics." (See also *photon*.)

quark. A hypothetical subnuclear particle having an electric charge one-third to two-thirds that of the electron. Also known as the fundamental subatomic particle, which is one of the smallest units of matter. (See also *subatomic particle*.)

radiation. The emission of energy in the form of electromagnetic waves, e.g., light (photons), radio waves, X-rays. Also used to denote the energy itself.

radioactive elements. Those elements subject to spontaneous disintegration (fission) of their atomic nuclei, with resultant emission of energy in the form of radiation. Radiation takes the form of alpha or beta particles or gamma rays or subatomic particles. (See also *radiation*.)

radiography. The process whereby an image is produced on a radiosensitive surface, i.e., photographic film, using radiation. For example, X-rays.

radioisotope. The isotopic form of a natural or synthetic element which exhibits radioactivity. The same as a radioactive isotope of an element.

rare-Earth elements. Elements found in the lanthanide series in Period 6, starting in Group 3 of the Periodic Table, beginning with atomic number 57 and ending with 71. They are usually found as oxides in minerals and ores. Some are produced as by-products from the radioactive decay of uranium and plutonium. They are not really "rare" or scarce but were just difficult to separate and identify.

reagent. A substance used in chemical reactions to detect, measure, examine, or analyze other substances.

rectification. The purification of vapor during the distillation process through interaction with a countercurrent liquid stream that is condensed from the vapor.

rectifier. A device, such as a diode, that converts alternating current to direct current.

reducing agent. A substance that induces reduction, i.e., acceptance of one or more electrons by an atom or ion, the removal of oxygen from a compound, or the addition of hydrogen to a compound. It is the opposite of oxidation, although the agent itself is oxidized (loses electrons).

refractory material. A material that is resistant to extremely high temperatures, e.g., carbon, chrome-ore magnesite, fire clay (aluminum silicates). Primarily used for lining steel furnaces and coke ovens.

salt. A compound formed when some or all of the hydrogen in an acid is replaced by a metal; e.g., the reaction of zinc with hydrochloric acid results in hydrogen gas and the salt zinc chloride ($2HCl + Zn \rightarrow ZnCl_2 + H_2$).

saturated compounds. Organic compounds containing only single bonds.

scintillation counter. See *Geiger counter*.

semiconductor. Usually a metalloid, such as silicon, which has conductive properties greater than those of an insulator but less than those of a conductor (metal).

It is possible to adjust their level of conductivity by changing the temperature or adding impurities. Semiconductors are important in the manufacture of computers and many other electronic devices.

shells. Theoretical spherical spaces surrounding the nucleus of an atom. They contain electrons. Also referred to as orbits or energy levels.

sintering. The process whereby powdered iron and other powdered metals or rare-Earths are "pressed" together to fit a mold, using minimum heat below the melting point of the materials. Sintering is used to produce homogeneous (uniform throughout) metal parts. It increases the strength, conductivity, and density of the product.

slag. The vitreous, i.e., glassy, mass or residue that is the leftover result of smelting of metallic ores.

smelting. Heating of an ore or mineral to separate the metallic component from the other substances in the ore by chemical reduction. Also known as "roasting."

solar evaporation. The evaporation of seawater by the sun in shallow ponds which results in the reclamation of sea salt and other compounds. The evaporated water, which is desalinated, can be condensed and collected as a source of drinking water.

solders. Alloys, usually containing lead and tin, which melt readily and are used to join metals.

solid. A state of matter in which the atoms or molecules of a substance are rigidly fixed in position. A solid has a fixed shape, a fixed mass, and a fixed volume.

solution. Usually considered a liquid mixture produced by dissolving a solid, liquid, or gas in a solvent. A solution is an even mixture at the molecular or ionic level of one or more substances. It can be a solution of solids as well as liquids or gases. Therefore, by definition, glass is a solution.

solvent. A substance that can dissolve another substance. Usually a liquid (the solvent) in which a solid, liquid, or gas (the solute) has dissolved to form a uniform mixture (the solution).

specific gravity. The density of a substance as related to the density of some standard, e.g.; water = 1. The specific gravity of solids and liquids is equivalent to their densities (g/cc). The specific gravity of gases, and to some extent liquids, is dependent on the temperature and pressure of the gas, because these factors affect the gas's volume. Therefore, the number representing the specific gravity of a gas may or may not be the same as the number for its density (g/L). (See also *density*.)

spectrograph. A spectroscope equipped to photograph spectra. (See *spectrum* for more detail.)

spectrometer. See *spectroscope*.

spectroscope. An instrument designed to analyze the wavelength of light emitted or absorbed by the excitation of elements. Each element, when heated, exhibits a unique line or color in the electromagnetic (visible light) spectrum.

spectroscopic analysis. See *spectroscopy*.

spectroscopy. The analysis of elements that separates the unique lightwaves either given off or absorbed by elements when heated.

spectrum. The range of all electromagnetic radiation, arranged progressively according to wavelength. The visible and near-visible spectrum of radiation ranges from the invisible rays called infrared, which we can't see but can feel as heat, to the range of visual light from deep red to violet, then to ultraviolet, whose wavelengths are too short to see but are just on the edge of the visual spectrum. (See also *electromagnetic spectrum.*)

steel. An alloy whose main constituent is iron. By adding different elements to form alloys of the iron, different steels can be produced, e.g., adding nickel to produce stainless steel.

subatomic particle. A component of an atom whose reactions are characteristic of the atom, e.g., electrons, protons, and neutrons, whereas subnuclear particles come from the nucleus during nuclear reactions and in cyclotrons, e.g., quarks, mesons, and so forth.

sublimation. The direct passage of a substance from a solid to a vapor without going through the liquid state.

superactinide elements. A proposed group of heavy, radioactive, synthetic, unstable elements, not yet produced in the laboratory. They are expected to have atomic numbers ranging from 119 to as high as 168 or 184, with very short half-lives, and existing in extremely small amounts.

superconductor. A metal, alloy, or compound which at temperatures near absolute zero loses both electrical resistance and magnetic permeability (is strongly repelled by magnets), thereby having infinite electrical conductivity.

supercooling. The cooling of a liquid below its normal freezing level without solidification.

synthesize. To create a new and complex product by combining separate elements.

synthetic. Man-made, i.e., not of natural origin.

tarnish. A type of corrosion caused by the reaction of a metal with hydrogen sulfide in the atmosphere, e.g., black silver sulfide forming on silver.

tensile. The ability to be stretched. Tensile strength is how much a material or object will resist being pulled apart.

tetravalent. Having a valence of 4.

Tevatron. A long tunnel through which scientists send high-speed protons or other subatomic particles. (See also *cyclotron.*)

thermochemical. A chemical reaction involving heat, e.g., endothermic (absorbs heat) and exothermic (gives off heat).

thermocouple. An instrument made up of dissimilar metals or semiconducting materials used to measure extremely high temperatures that are beyond the range of liquid-in-glass thermometers. Also, a bimetal used in home thermostats to control heat and cooling furnaces and air conditioners.

thermoelectrical. The characteristic of electricity generated by a temperature difference between two dissimilar materials. Usually semiconductors or metallic conductors. (See also *thermocouple*.)

thermonuclear reaction. Release of heat energy when the nuclei of atoms split (fission, atom bomb, or nuclear power plant) or when nuclei combine (fusion, hydrogen bomb).

toxic. Damaging or harmful to the body or its organs and cells, e.g., poison.

tracer. A radioactive entity, e.g., radioisotope, which is added to the reacting elements and/or compounds in a chemical process and which can be traced and detected through that process. Some examples of tracers are carbon-14 and, in medicine, radioactive forms of iodine and sodium.

transactinide elements. A number of unstable radioactive elements that extend beyond the transuranic elements, i.e., beyond lawrencium, with atomic numbers of 104 through 112. They do not exist in nature and have very short half-lives; some exist for only a fraction of a second and are synthesized in very small amounts.

transistor. A device that overcomes the resistance when a current of electricity passes through it. Transistors are important in the manufacture of many types of electronics.

transition elements. Elements located in Periods 4 and 5, ranging from Group 3 of the Periodic Table through Group 12. They are metals that represent a gradual shift from those having a very weak electronegativity (form positive ions) to the strongly electronegative nonmetallic elements that form negative ions. That is, they are elements in "transition" from metals to nonmetals. The lanthanide and actinide series of elements are also considered "transition" elements because they use electron shells inside the outer shell as their valence electron shells and are classed as metals.

transmutation. The transformation, either artificially or naturally, of the atoms of one element into the atoms of a different element as the result of nuclear reactions.

transurancic element. A radioactive element of the actinide series with a higher atomic number than uranium (U^{92}). They are radioactive and produced by nuclear bombardment.

triatomic molecule. A molecule that contains three atoms, e.g., ozone (O_3) or magnesium chloride ($MgCl_2$).

trivalent. Having a valence of 3. For example, elements in Group 13 of the Periodic Table.

unsaturated. Compounds, especially carbon, containing atoms that share more than one valence bond and are capable of dissolving more of a solute at a given temperature.

ultraviolet (UV). The radiation wavelength in the electromagnetic spectrum from 100 to 3900 angstroms (Å), between the X-ray region and visible violet light.

vacuum. A space devoid of matter, i.e., containing no atoms or molecules.

valence. The whole number that represents the combining power of one element with another element. Valence electrons are usually, but not always, the electrons in the outermost shell.

vasodilator. A chemical agent that causes the dilation of blood vessels.

viscosity. The internal resistance to which a liquid resists flowing when pressure is applied. The lower the viscosity, the easier the substance will flow.

vulcanization. A process in which sulfur is added to rubber, resulting in the increased hardness of the rubber.

water. A colorless liquid, which is a compound consisting of the atoms of hydrogen and oxygen. It is found in the atmosphere, in the Earth, and in all living creatures. It is essential to sustain life. Pure water has a pH of 7 (neutral).

X-ray diffraction. A method of spectroanalysis of substances using X-rays instead of light waves. It involves scattering of the X-radiation at angles that are unique to the substance being analyzed, thus yielding a spectrum of the atomic or molecular substance being analyzed.

ALPHABETICAL INDEX OF THE ELEMENTS

Name of Element	Chemical Symbol	Atomic Number	Atomic Weight	Page Number
Actinium	Ac	89	227.028	246
Aluminum	Al	13	26.9815	139
Americium	Am	95	243	256
Antimony	Sb	51	121.75	171
Argon	Ar	18	39.948	212
Arsenic	As	33	74.9216	169
Astatine	At	85	210	204
Barium	Ba	56	137.34	65
Berkelium	Bk	97	247	258
Beryllium	Be	4	9.0122	56
Bismuth	Bi	83	208.98	173
Boron	B	5	10.81	136
Bromine	Br	35	79.91	200
Cadmium	Cd	48	112.41	113
Calcium	Ca	20	40.08	60
Californium	Cf	98	252	260
Carbon	C	6	12.011	147

Name of Element	Chemical Symbol	Atomic Number	Atomic Weight	Page Number
Cerium	Ce	58	140.12	223
Cesium	Cs	55	132.905	51
Chlorine	Cl	17	35.453	198
Chromium	Cr	24	51.996	79
Cobalt	Co	27	58.933	87
Copper	Cu	29	63.546	91
Curium	Cm	96	244	257
Dysprosium	Dy	66	162.50	234
Einsteinium	Es	99	254	261
Erbium	Er	68	167.26	237
Europium	Eu	63	151.96	230
Fermium	Fm	100	252	262
Fluorine	F	9	18.998	195
Francium	Fr	87	223	54
Gadolinium	Gd	64	157.25	231
Gallium	Ga	31	69.72	142
Germanium	Ge	32	72.59	152
Gold	Au	79	196.967	127
Hafnium	Hf	72	178.49	115
Helium	He	2	4.0026	207
Holmium	Ho	67	164.93	234
Hydrogen	H	1	1.0079	37
Indium	In	49	114.82	143
Iodine	I	53	126.905	202
Iridium	Ir	77	192.22	123
Iron	Fe	26	55.847	84
Krypton	Kr	36	83.80	213
Lanthanum	La	57	138.906	221
Lawrencium	Lr (Lw)	103	260	267
Lead	Pb	82	207.19	157
Lithium	Li	3	6.941	41
Lutetium	Lu	71	174.967	241
Magnesium	Mg	12	24.305	58

Name of Element	Chemical Symbol	Atomic Number	Atomic Weight	Page Number
Manganese	Mn	25	54.938	81
Mendelevium	Md	101	256	264
Mercury	Hg	80	200.59	129
Molybdenum	Mo	42	95.94	102
Neodymium	Nd	60	144.24	226
Neon	Ne	10	20.179	210
Neptunium	Np	93	237	252
Nickel	Ni	28	58.699	89
Niobium	Nb	41	92.906	100
Nitrogen	N	7	14.0067	163
Nobelium	No	102	254	265
Osmium	Os	76	190.2	121
Oxygen	O	8	15.9994	175
Palladium	Pd	46	106.42	109
Phosphorus	P	15	30.9738	166
Platinum	Pt	78	195.09	124
Plutonium	Pu	94	244	254
Polonium	Po	84	210	190
Potassium	K	19	39.0983	46
Praseodymium	Pr	59	140.908	224
Promethium	Pm	61	145	227
Protactinium	Pa	91	231	249
Radium	Ra	88	226	67
Radon	Rn	86	222	216
Rhenium	Re	75	186.207	120
Rhodium	Rh	45	102.906	108
Rubidium	Rb	37	85.4778	49
Ruthenium	Ru	44	101.07	106
Samarium	Sm	62	150.36	229
Scandium	Sc	21	44.9559	73
Selenium	Se	34	78.96	186
Silicon	Si	14	28.0855	150
Silver	Ag	47	107.868	111

Name of Element	Chemical Symbol	Atomic Number	Atomic Weight	Page Number
Sodium	Na	11	22.9898	44
Strontium	Sr	38	87.62	63
Sulfur	S	16	32.066	184
Tantalum	Ta	73	180.948	116
Technetium	Tc	43	99	104
Tellurium	Te	52	127.60	188
Terbium	Tb	65	158.925	233
Thallium	Tl	81	204.383	145
Thorium	Th	90	232.038	248
Thulium	Tm	69	168.934	238
Tin	Sn	50	118.71	154
Titanium	Ti	22	47.88	75
Tungsten	W	74	183.85	118
Uranium	U	92	238.029	250
Vanadium	V	23	50.9415	77
Xenon	Xe	54	131.29	214
Ytterbium	Yb	70	173.04	239
Yttrium	Y	39	88.9050	96
Zinc	Zn	30	65.39	93
Zirconium	Zr	40	91.224	98

Transactinide elements beyond lawrencium 103. All are artificially made in very small amounts, all are unstable, radioactive, and have very short half-lives.

Unnilquadium	Unq	104	261	271
Unnilpentium	Unp	105	262	272
Unnilhexium	Unh	106	263	274
Unnilseptium	Uns	107	264	275
Unniloctium	Uno	108	265	276
Unnilennium	Une	109	266	277
Ununnilium	Uun	110	267?	277
Unununium	Uuu	111	272	279
Ununbiium	Uub	112	277	280

(See Chapter 10 for more details on these newer man-made elements.)

ELECTRON CONFIGURATION OF THE ATOMS

MAXIMUM NUMBER OF ELECTRONS PER SHELL AND SUBORBITALS

K Shell: 2 Electrons;   (Suborbital #1: s2)

L Shell: 8 Electrons;   (Suborbitals #2: s2 + p6)

M Shell: 18 Electrons;   (Suborbitals #3: s2 + p6 + d10)

N Shell: 32 Electrons;   (Suborbitals #4: s2 + p6 + d10 + f14)

O Shell: 32 Electrons;   (Suborbitals #5: s2 + p6 + d10 + f14)

P Shell: 18 Electrons;   (Suborbitals #6: s2 + p6 + d10)

Q Shell: 2 Electrons;   (Suborbital #7: s2)

Name of Element	Chemical Symbol	Atomic Number	Electron Shells	Electron Suborbitals
Chapter 4—Group 1: Alkali Metals				
Hydrogen	H	1	SHELLS K = 1	Suborbitals s1
Lithium	Li	3	K = 2 L = 1	s2 p1

Name of Element	Chemical Symbol	Atomic Number	Electron Shells	Electron Suborbitals
Sodium	Na	11	SHELLS K = 2 L = 8 M = 1	Suborbitals s2 s2, p6 s1
Potassium	K	19	SHELLS K = 2 L = 8 M = 8 N = 1	Suborbitals s2 s2, p6 s2, p6 s1
Rubidium	Rb	37	SHELLS K = 2 L = 8 M = 18 N = 8 O = 1	Suborbitals s2 s2, p6 s2, p6, d10 s2, p6 s1
Cesium	Cs	55	SHELLS K = 2 L = 8 M = 18 N = 18 O = 8 P = 1	Suborbitals s2 s2, p6 s2, p6, d10 s2, p6, d10 s2, p6 s1
Francium	Fr	87	SHELLS K = 2 L = 8 M = 18 N = 32 O = 18 P = 8 Q = 1	Suborbitals s2 s2, p6 s2, p6, d10 s2, p6, d10, f14 s2, p6, d10 s2, p6 s1

Name of Element	Chemical Symbol	Atomic Number	Electron Shells	Electron Suborbitals

Chapter 4—Group 2: Alkali Earth Metals

Name of Element	Chemical Symbol	Atomic Number	Electron Shells	Electron Suborbitals
Beryllium	Be	4	SHELLS	Suborbitals
			K = 2	s2
			L = 2	s2
Magnesium	Mg	12	SHELLS	Suborbitals
			K = 2	s2
			L = 8	s2, p6
			M = 2	s2
Calcium	Ca	20	SHELLS	Suborbitals
			K = 2	s2
			L = 8	s2, p6
			M = 8	s2, p6
			N = 2	s2
Strontium	Sr	38	SHELLS	Suborbitals
			K = 2	s2
			L = 8	s2, p6
			M = 18	s2, p6, d10
			N = 8	s2, p6
			O = 2	s2
Barium	Ba	56	SHELLS	Suborbitals
			K = 2	s2
			L = 8	s2, p6
			M = 18	s2, p6, d10
			N = 18	s2, p6, d10
			O = 8	s2, p6
			P = 2	s2
Radium	Ra	88	SHELLS	Suborbitals
			K = 2	s2
			L = 8	s2, p6
			M = 18	s2, p6, d10
			N = 18	s2, p6, d10

Name of Element	Chemical Symbol	Atomic Number	Electron Shells	Electron Suborbitals
			O = 32	s2, p6, d10, f14
			P = 8	s2, p6
			Q = 2	s2

Chapter 5—Transition Elements: First Series
Period 4; Starting at Group 3 (IIIB)

Name of Element	Chemical Symbol	Atomic Number	Electron Shells	Electron Suborbitals
Scandium	Sc	21	SHELLS	Suborbitals
			K = 2	s2
			L = 8	s2, p6
			M = 9	s2, p6, d1
			N = 2	s2
Titanium	Ti	22	SHELLS	Suborbitals
			K = 2	s2
			L = 8	s2, p6
			M = 10	s2, p6, d2
			N = 2	s2
Vanadium	V	23	SHELLS	Suborbitals
			K = 2	s2
			L = 8	s2, p6
			M = 11	s2, p6, d3
			N = 2	s2
Chromium	Cr	24	SHELLS	Suborbitals
			K = 2	s2
			L = 8	s2, p6
			M = 13	s2, p6, d5
			N = 1	s1
Manganese	Mn	25	SHELLS	Suborbitals
			K = 2	s2
			L = 8	s2, p6
			M = 13	s2, p6, d5
			N = 2	s2

Name of Element	Chemical Symbol	Atomic Number	Electron Shells	Electron Suborbitals
Iron	Fe	26	SHELLS	Suborbitals
			K = 2	s2
			L = 8	s2, p6
			M = 14	s2, p6, d6
			N = 2	s2
Cobalt	Co	27	SHELLS	Suborbitals
			K = 2	s2
			L = 8	s2, p6
			M = 15	s2, p6, d7
			N = 2	s2
Nickel	Ni	28	SHELLS	Suborbitals
			K = 2	s2
			L = 8	s2, p6
			M = 16	s2, p6, d8
			N = 2	s2
Copper	Cu	29	SHELLS	Suborbitals
			K = 2	s2
			L = 8	s2, p6
			M = 18	s2, p6, d10
			N = 1	s1
Zinc	Zn	30	SHELLS	Suborbitals
			K = 2	s2
			L = 8	s2, p6
			M = 18	s2, p6, d10
			N = 2	s2

Chapter 5—Transition Elements: Second Series
Period 5; Starting at Group 3 (IIIB)

Yttrium	Y	39	SHELLS	Suborbitals
			K = 2	s2
			L = 8	s2, p6

Name of Element	Chemical Symbol	Atomic Number	Electron Shells	Electron Suborbitals
			M = 18	s2, p6, d10
			N = 9	S2, P6, d1
			O = 2	S2
Zirconium	Zr	40	SHELLS	Suborbitals
			K = 2	s2
			L = 8	s2, p6
			M = 18	s2, p6, d10
			N = 10	s2, p6, d2
			O = 2	s2
Niobium	Nb	41	SHELLS	Suborbitals
			K = 2	s2
			L = 8	s2, p6
			M = 18	s2, p6, d10
			N = 12	s2, p6, d4
			O = 1	s1
Molybdenum	Mo	42	SHELLS	Suborbitals
			K = 2	s2
			L = 8	s2, p6
			M = 18	s2, p6, d10
			N = 13	s2, p6, d5
			O = 1	s1
Technetium	Tc	43	SHELLS	Suborbitals
			K = 2	s2
			L = 8	s2, p6
			M = 18	s2, p6, 10
			N = 13	s2, p6, d5
			O = 2	s2
Ruthenium	Ru	44	SHELLS	Suborbitals
			K = 2	s2
			L = 8	s2, p6

Name of Element	Chemical Symbol	Atomic Number	Electron Shells	Electron Suborbitals
			M = 18	s2, p6, d10
			N = 15	s2, p6, d7
			O = 1	s1
Rhodium	Rh	45	SHELLS	Suborbitals
			K = 2	s2
			L = 8	s2, p6
			M = 18	s2, p6, d10
			N = 16	s2, p6, d8
			O = 1	s1
Palladium	Pd	46	SHELLS	Suborbitals
			K = 2	s2
			L = 8	s2, p6
			M = 18	s2, p6, d10
			N = 18	s2, p6, d10
			O = 0	s0
Silver	Ag	47	SHELLS	Suborbitals
			K = 2	s2
			L = 8	s2, p6
			M = 18	s2, p6, d10
			N = 18	s2, p6, d10
			O = 1	s1
Cadmium	Cd	48	SHELLS	Suborbitals
			K = 2	s2
			L = 8	s2, p6
			M = 18	s2, p6, d10
			N = 18	s2, p6, d10
			O = 2	s2

Name of Element	Chemical Symbol	Atomic Number	Electron Shells	Electron Suborbitals

Chapter 5—Transition Elements: Third Series
Period 6; Following the Lanthanide Series

Hafnium	Hf	72	SHELLS	Suborbitals
			K = 2	s2
			L = 8	s2, p6
			M = 18	s2, p6, d10
			N = 32	s2, p6, d10, f14
			O = 10	s2, p6, d2
			P = 2	s2
Tantalum	Ta	73	SHELLS	Suborbitals
			K = 2	s2
			L = 8	s2, p6
			M = 18	s2, p6, d10
			N = 32	s2, p6, d10, f14
			O = 11	s2, p6, d3
			P = 2	s2
Tungsten	W	74	SHELLS	Suborbitals
			K = 2	s2
			L = 8	s2, p6
			M = 18	s2, p6, d10
			N = 32	s2, p6, d10, f14
			O = 12	s2, p6, d4
			P = 2	s2
Rhenium	Re	75	SHELLS	Suborbitals
			K = 2	s2
			L = 8	s2, p6
			M = 18	s2, p6, d10
			N = 32	s2, p6, d10, f14
			O = 13	s2, p6, d5
			P = 2	s2

Name of Element	Chemical Symbol	Atomic Number	Electron Shells	Electron Suborbitals
Osmium	Os	76	SHELLS K = 2 L = 8 M = 18 N = 32 O = 14 P = 2	Suborbitals s2 s2, p6 s2, p6, d10 s2, p6, d10, f14 s2, p6, d6 s2
Iridium	Ir	77	SHELLS K = 2 L = 8 M = 18 N = 32 O = 15 P = 2	Suborbitals s2 s2, p6 s2, p6, d10 s2, p6, d10, f14 s2, p6, d7 s2
Platinum	Pt	78	SHELLS K = 2 L = 8 M = 18 N = 32 O = 17 P = 1	Suborbitals s2 s2, p6 s2, p6, d10 s2, p6, d10, f14 s2, p6, d9 s1
Gold	Au	79	SHELLS K = 2 L = 8 M = 18 N = 32 O = 18 P = 1	Suborbitals s2 s2, p6 s2, p6, d10 s2, p6, d10, f14 s2, p6, d10 s1
Mercury	Hg	80	SHELLS K = 2 L = 8	Suborbitals s2 s2, p6

Name of Element	Chemical Symbol	Atomic Number	Electron Shells	Electron Suborbitals
			M = 18	s2, p6, d10
			N = 32	s2, p6, d10, f14
			O = 18	s2, p6, d10
			P = 2	s2

Chapter 6—Group 13: Metallics

Name of Element	Chemical Symbol	Atomic Number	Electron Shells	Electron Suborbitals
Boron	B	5	SHELLS	Suborbitals
			K = 2	s2
			L = 3	s2, p1
Aluminum	Al	13	SHELLS	Suborbitals
			K = 2	s2
			L = 8	s2, p6
			M = 3	s2, p1
Gallium	Ga	31	SHELLS	Suborbitals
			K = 2	s2
			L = 8	s2, p6
			M = 18	s2, p6, d10
			N = 3	s2, p1
Indium	In	49	SHELLS	Suborbitals
			K = 2	s2
			L = 8	s2, p6
			M = 18	s2, p6, d10
			N = 18	s2, p6, d10
			O = 3	s2, p1
Thallium	Tl	81	SHELLS	Suborbitals
			K = 2	s2
			L = 8	s2, p6
			M = 18	s2, p6, d10
			N = 32	s2, p6, d10, f14
			O = 18	s2, p6, d10
			P = 3	s2, p1

Name of Element	Chemical Symbol	Atomic Number	Electron Shells	Electron Suborbitals
Chapter 6—Group 14: Metalloids				
Carbon	C	6	SHELLS	Suborbitals
			K = 2	s2
			L = 4	s2, p2
Silicon	Si	14	SHELLS	Suborbitals
			K = 2	s2
			L = 8	s2, p6
			M = 4	s2, p2
Germanium	Ge	32	SHELLS	Suborbitals
			K = 2	s2
			L = 8	s2, p6
			M = 18	s2, p6, d10
			N = 4	s2, p2
Tin	Sn	50	SHELLS	Suborbitals
			K = 2	s2
			L = 8	s2, p6
			M = 18	s2, p6, d10
			N = 18	s2, p6, d10
			O = 4	s2, p2
Lead	Pb	82	SHELLS	Suborbitals
			K = 2	s2
			L = 8	s2, p6
			M = 18	s2, p6, d10
			N = 32	s2, p6, d10, f14
			O = 18	s2, p6, d10
			P = 4	s2, p2
Chapter 7—Group 15: Metalloids				
Nitrogen	N	7	SHELLS	Suborbitals
			K = 2	s2
			L = 5	s2, p3

Name of Element	Chemical Symbol	Atomic Number	Electron Shells	Electron Suborbitals
Phosphorus	P	15	SHELLS	Suborbitals
			K = 2	s2
			L = 8	s2, p6
			M = 5	s2, p3
Arsenic	As	33	SHELLS	Suborbitals
			K = 2	s2
			L = 8	s2, p6
			M = 18	s2, p6, d10
			N = 5	s2, p3
Antimony	Sb	51	SHELLS	Suborbitals
			K = 2	s2
			L = 8	s2, p6
			M = 18	s2, p6, d10
			N = 18	s2, p6, d10
			O = 5	s2, p3
Bismuth	Bi	83	SHELLS	Suborbitals
			K = 2	s2
			L = 8	s2, p6
			M = 18	s2, p6, d10
			N = 32	s2, p6, d10, f14
			O = 18	s2, p6, d10
			P = 5	s2, p3

Chapter 7—Group 16: Nonmetals

Name of Element	Chemical Symbol	Atomic Number	Electron Shells	Electron Suborbitals
Oxygen	O	8	SHELLS	Suborbitals
			K = 2	s2
			L = 6	s2, p4
Sulfur	S	16	SHELLS	Suborbitals
			K = 2	s2
			L = 8	s2, p6
			M = 6	s2, p4

Name of Element	Chemical Symbol	Atomic Number	Electron Shells	Electron Suborbitals
Selenium	Se	34	SHELLS K = 2 L = 8 M = 18 L = 6	Suborbitals s2 s2, p6 s2, p6, d10 s2, p4
Tellurium	Te	52	SHELLS K = 2 L = 8 M = 18 N = 18 O = 6	Suborbitals s2 s2, p6 s2, p6, d10 s2, p6, d10 s2, p4
Polonium	Po	84	SHELLS K = 2 L = 8 M = 18 N = 32 O = 18 P = 6	Suborbitals s2 s2, p6 s2, p6, d10 s2, p6, d10, f14 s2, p6, d10 s2, p4

Chapter 8—Group 17: The Halogens

Name of Element	Chemical Symbol	Atomic Number	Electron Shells	Electron Suborbitals
Fluorine	F	9	SHELLS K = 2 L = 7	Suborbitals s2 s2, p5
Chlorine	Cl	17	SHELLS K = 2 L = 8 N = 7	Suborbitals s2 s2, p6 s2, p5
Bromine	Br	35	SHELLS K = 2 L = 8	Suborbitals s2 s2, p6

Name of Element	Chemical Symbol	Atomic Number	Electron Shells	Electron Suborbitals
			M = 18	s2, p6, d10
			N = 7	s2, p5
Iodine	I	53	SHELLS	Suborbitals
			K = 2	s2
			L = 8	s2, p6
			M = 18	s2, p6, d10
			N = 18	s2, p6, d10
			O = 7	s2, p5
Astatine	At	85	SHELLS	Suborbitals
			K = 2	s2
			L = 8	s2, p6
			M = 18	s2, p6, d10
			N = 32	s2, p6, d10, f14
			O = 18	s2, p6, d10
			P = 7	s2, p5

Chapter 8—Group 18: Noble, Inert Gases

Name of Element	Chemical Symbol	Atomic Number	Electron Shells	Electron Suborbitals
Helium	He	2	SHELLS	Suborbitals
			K = 2	s2
Neon	Ne	10	SHELLS	Suborbitals
			K = 2	s2
			L = 8	s2, p6
Argon	Ar	18	SHELLS	Suborbitals
			K = 2	s2
			L = 8	s2, p6
			M = 8	s2, p6
Krypton	Kr	36	SHELLS	Suborbitals
			K = 2	s2
			L = 8	s2, p6
			M = 18	s2, p6, d10
			N = 8	s2, p6

Name of Element	Chemical Symbol	Atomic Number	Electron Shells	Electron Suborbitals
Xenon	Xe	54	SHELLS K = 2 L = 8 M = 18 N = 18 O = 8	Suborbitals s2 s2, p6 s2, p6, d10 s2, p6, d10 s2, p6
Radon	Rn	86	SHELLS K = 2 L = 8 M = 18 N = 32 O = 18 P = 8	Suborbitals s2 s2, p6 s2, p6, d10 s2, p6, d10, f14 s2, p6, d10 s2, p6

Chapter 9—Lanthanide Series: Rare-Earths
Period 6, Starting at Group 3 (IIIB)

Name of Element	Chemical Symbol	Atomic Number	Electron Shells	Electron Suborbitals
Lanthanum	La	57	SHELLS K = 2 L = 8 M = 18 N = 18 O = 9 P = 2	Suborbitals s2 s2, p6 s2, p6, d10 s2, p6, d10 s2, p6, d1 s2
Cerium	Ce	58	SHELLS K = 2 L = 8 M = 18 N = 20 O = 8 P = 2	Suborbitals s2 s2, p6 s2, p6, d10 s2, p6, d10, f2 s2, p6 s2
Praseodymium	Pr	59	SHELLS K = 2 L = 8	Suborbitals s2 s2, p6

Name of Element	Chemical Symbol	Atomic Number	Electron Shells	Electron Suborbitals
			M = 18	s2, p6, d10
			N = 21	s2, p6, d10, f3
			O = 8	s2, p6
			P = 2	s2
Neodymium	Nd	60	SHELLS	Suborbitals
			K = 2	s2
			L = 8	s2, p6
			M = 18	s2, p6, d10
			N = 22	s2, p6, d10, f4
			O = 8	s2, p6
			P = 2	s2
Promethium	Pm	61	SHELLS	Suborbitals
			K = 2	s2
			L = 8	s2, p6
			M = 18	s2, p6, d10
			N = 23	s2, p6, d10, f5
			O = 8	s2, p6
			P = 2	s2
Samarium	Sm	62	SHELLS	Suborbitals
			K = 2	s2
			L = 8	s2, p6
			M = 18	s2, p6, d10
			N = 24	s2, p6, d10, f6
			O = 8	s2, p6
			P = 2	s2
Europium	Eu	63	SHELLS	Suborbitals
			K = 2	s2
			L = 8	s2, p6
			M = 18	s2, p6, d10
			N = 25	s2, p6, d10, f7

Name of Element	Chemical Symbol	Atomic Number	Electron Shells	Electron Suborbitals
			O = 8	s2, p6
			P = 2	s2
Gadolinum	Gd	64	SHELLS	Suborbitals
			K = 2	s2
			L = 8	s2, p6
			M = 18	s2, p6, d10
			N = 25	s2, p6, d10, f7
			O = 9	s2, p6, d1
			P = 2	s2
Terbium	Tb	65	SHELLS	Suborbitals
			K = 2	s2
			L = 8	s2, p6
			M = 18	s2, p6, d10
			N = 27	s2, p6, d10, f9
			O = 8	s2, p6
			P = 2	s2
Dysprosium	Dy	66	SHELLS	Suborbitals
			K = 2	s2
			L = 8	s2, p6
			M = 18	s2, p6, d10
			N = 28	s2, p6, d10, f10
			O = 8	s2, p6
			P = 2	s2
Holmium	Ho	67	SHELLS	Suborbitals
			K = 2	s2
			L = 8	s2, p6
			M = 18	s2, p6, d10
			N = 29	s2, p6, d10, f11
			O = 8	s2, p6
			P = 2	s2

Name of Element	Chemical Symbol	Atomic Number	Electron Shells	Electron Suborbitals
Erbium	Er	68	SHELLS	Suborbitals
			K = 2	s2
			L = 8	s2, p6
			M = 18	s2, p6, d10
			N = 30	s2, p6, d10, f12
			O = 8	s2, p6
			P = 2	s2
Thulium	Tm	69	SHELLS	Suborbitals
			K = 2	s2
			L = 8	s2, p6
			M = 18	s2, p6, d10
			N = 31	s2, p6, d10, f13
			O = 8	s2, p6
			P = 2	s2
Ytterbium	Yb	70	SHELLS	Suborbitals
			K = 2	s2
			L = 8	s2, p6
			M = 18	s2, p6, d10
			N = 32	s2, p6, d10, f14
			O = 8	s2, p6
			P = 2	s2
Lutetium	Lu	71	SHELLS	Suborbitals
			K = 2	s2
			L = 8	s2, p6
			M = 18	s2, p6, d10
			N = 32	s2, p6, d10, f14
			O = 9	s2, p6, d1
			P = 2	s2

Name of Element	Chemical Symbol	Atomic Number	Electron Shells	Electron Suborbitals

Chapter 9—Actinide Series: Including the Transuranic Elements—Following U^{92}

Period 7; Starting at Group 3 (IIIB)

Actinium	Ac	89	SHELLS	Suborbitals
			K = 2	s2
			L = 8	s2, p6
			M = 18	s2, p6, d10
			N = 32	s2, p6, d10, f14
			O = 18	s2, p6, d10
			P = 9	s2, p6, d1
			Q = 2	s2
Thorium	Th	90	SHELLS	Suborbitals
			K = 2	s2
			L = 8	s2, p6
			M = 18	s2, p6, d10
			N = 32	s2, p6, d10, f14
			O = 18	s2, p6, d10
			P = 10	s2, p6, d2
			Q = 2	s2
Protactinium	Pa	91	SHELLS	Suborbitals
			K = 2	s2
			L = 8	s2, p6
			M = 18	s2, p6, d10
			N = 32	s2, p6, d10, f14
			O = 20	s2, p6, d10, f2
			P = 9	s2, p6, d1
			Q = 2	s2
Uranium	U	92	SHELLS	Suborbitals
			K = 2	s2
			L = 8	s2, p6
			M = 18	s2, p6, d10
			N = 32	s2, p6, d10, f14
			O = 21	s2, p6, d10, f3

Name of Element	Chemical Symbol	Atomic Number	Electron Shells	Electron Suborbitals
			P = 9	s2, p6, d1
			Q = 2	s2
Neptunium	Np	93	SHELLS	Suborbitals
			K = 2	s2
			L = 8	s2, p6
			M = 18	s2, p6, d10
			N = 32	s2, p6, d10, f14
			O = 22	s2, p6, d10, f4
			P = 9	s2, p6, d1
			Q = 2	s2
Plutonium	Pu	94	SHELLS	Suborbitals
			K = 2	s2
			L = 8	s2, p6
			M = 18	s2, p6, d10
			N = 32	s2, p6, d10, f14
			O = 23	s2, p6, d10, f5
			P = 9	s2, p6, d1
			Q = 2	s2
Americium	Am	95	SHELLS	Suborbitals
			K = 2	s2
			L = 8	s2, p6
			M = 18	s2, p6, d10
			N = 32	s2, p6, d10, f14
			O = 24	s2, p6, d10, f6
			P = 9	s2, p6, d1
			Q = 2	s2
Curium	Cm	96	SHELLS	Suborbitals
			K = 2	s2
			L = 8	s2, p6
			M = 18	s2, p6, d10

Name of Element	Chemical Symbol	Atomic Number	Electron Shells	Electron Suborbitals
			N = 32	s2, p6, d10, f14
			O = 25	s2, p6, d10, f7
			P = 9	s2, p6, d1
			Q = 2	s2
Berkelium	Bk	97	SHELLS	Suborbitals
			K = 2	s2
			L = 8	s2, p6
			M = 18	s2, p6, d10
			N = 32	s2, p6, d10, f14
			O = 26	s2, p6, d10, f8
			P = 9	s2, p6, d1
			Q = 2	s2
Californium	Cf	98	SHELLS	Suborbitals
			K = 2	s2
			L = 8	s2, p6
			M = 18	s2, p6, d10
			N = 32	s2, p6, d10, f14
			O = 27	s2, p6, d10, f9
			P = 9	s2, p6, d1
			Q = 2	s2
Einsteinium	Es	99	SHELLS	Suborbitals
			K = 2	s2
			L = 8	s2, p6
			M = 18	s2, p6, d10
			N = 32	s2, p6, d10, f14
			O = 28	s2, p6, d10, f10
			P = 9	s2, p6, d1
			Q = 2	s2
Fermium	Fm	100	SHELLS	Suborbitals
			K = 2	s2
			L = 8	s2, p6

Name of Element	Chemical Symbol	Atomic Number	Electron Shells	Electron Suborbitals
			M = 18	s2, p6, d10
			N = 32	s2, p6, d10, f14
			O = 29	s2, p6, d10, f11
			P = 9	s2, p6, d1
			Q = 2	s2
Mendelevium	Md	101	SHELLS	Suborbitals
			K = 2	s2
			L = 8	s2, p6
			M = 18	s2, p6, d10
			N = 32	s2, p6, d10, f14
			O = 30	s2, p6, d10, f12
			P = 9	s2, p6, d1
			Q = 2	s2
Nobelium	No	102	SHELLS	Suborbitals
			K = 2	s2
			L = 8	s2, p6
			M = 18	s2, p6, d10
			N = 32	s2, p6, d10, f14
			O = 31	s2, p6, d10, f13
			P = 9	s2, p6, d1
			Q = 2	s2
Lawrencium	Lr (Lw)	103	SHELLS	Suborbitals
			K = 2	s2
			L = 8	s2, p6
			M = 18	s2, p6, d10
			N = 32	s2, p6, d10, f14
			O = 32	s2, p6, d10, f14
			P = 9	s2, p6, d1
			Q = 2	s2

SELECTED BIBLIOGRAPHY

Asimov, Isaac. *Asimov's Biographical Encyclopedia of Science and Technology.* New York: Doubleday & Company, 1964.

———. *Asimov's Chronology of Science and Discovery.* New York: Harper & Row, 1989.

———. *Beginnings, The Story of Origins—of Mankind, Life, the Earth, the Universe.* New York: A Berkeley Book, 1989.

Atomic Structure. (Chemistry Concepts On-Line Series for Grades 9–12, MS-DOS). Thornwood, NY: Edunetics, 1996.

Banks, Alton J. and Holmes, Jon L. *Periodic Table CD, Journal of Chemical Education Software.* Madison, WI: University of Wisconsin-Madison, 1995.

Barnes-Svarney, Patricia. Editorial Director. *The New York Public Library Science Desk Reference.* New York: Simon & Schuster, Macmillan, 1995.

Boorstin, Daniel J. *The Discoverers.* New York: Vintage Books, 1983.

Campbell, Norman. *What Is Science?* New York: Dover Publications, 1953.

Carbon Cycle, The. (DOS Software for Windows). Redmond, WA: Software Labs, 1992.

Chemical Heritage Foundation. Philadelphia: The Beckman Center for History of Chemistry. (A reference source for chemical history.)

Chemicals for Windows. (DOS Software for Windows). Redmond, WA: Software Labs, 1992.

Chemistry. (High School Level Interactive Multimedia CD Software for Windows). San Jose, CA: Sequoia Business Co., 1994.

Chemistry. (Interactive CD for Windows). Saratoga, CA: Stanford Studyware, 1995.

Complete Home Medical Guide, The. The Columbia University College of Physicians and Surgeons. New York: Crown Publishers, 1985.

Cromer, Alan. *Uncommon Sense.* New York: Oxford University Press, 1993.

Elements. (Science and Nature Interactive CD-ROM Software Series). London: Mentorom Multimedia, YITM, Ltd., 1995.

Ferris, Timothy. *Coming of Age in the Milky Way.* New York: Doubleday & Company, 1988.

Gamow, George. *Mr. Tompkins in Paperback.* Cambridge: Cambridge University Press, 1994.

Handy Science Answer Book, The. Carnegie Library of Pittsburgh. Detroit: Visible Ink Press, 1994.

Holmyard, E. J. *Alchemy.* New York: Dover Publications, 1990. (First published in Middlesex, England by Penguin Books, 1957.)

Ihde, Aaron J. *The Development of Modern Chemistry.* New York: Dover Publications, 1984.

Jaffe, Bernard. *Crucibles: The Story of Chemistry.* New York: Dover Publications, 1976 (Simon & Schuster, 1930).

Lewis, Richard J. *Hawley's Condensed Chemical Dictionary.* 12th ed. New York: Van Nostrand Reinhold, 1993.

Mclellan, C. R., Day, Marion C., and Clark, Roy W. *Concepts of General Chemistry.* Philadelphia: F. A. Davis Company, 1966.

Name That Element. (Computer graphic games I & II, DOS). Redmond, WA: Software Labs, 1992.

Pauling, Linus. *General Chemistry.* New York: Dover Publications, 1970.

Periodic Table and Chemical Formulas, The. (Chemistry Concepts On-Line for Grades 9–12, DOS). Thornwood, NY: Edunetics, 1996.

Read, John. *From Alchemy to Chemistry.* Toronto, Ontario: General Publishing Co., 1995.

Schneider, Herman and Schneider, Leo. *The Harper Dictionary of Science in Everyday Language.* New York: Harper & Row, 1988.

Trefil, James. *1001 Things Everyone Should Know about Science.* New York: Doubleday & Company, 1992.

Way Science Works, The. Macmillan. New York: Simon & Schuster, Macmillan, 1995.

INDEX

About the Author

ROBERT E. KREBS is the retired Associate Dean for Research at the University of Illinois Health Sciences Center, Chicago. Since his retirement, Dr. Krebs has authored over 30 computer-based instructional courseware programs in areas covering science education, health/medical practices, and academic administration.